MATHS PLUS 3

STAGE 2

MENTALS AND HOMEWORK BOOK

NEW SOUTH WALES SYLLABUS

Harry O'Brien
Greg Purcell

OXFORD
UNIVERSITY PRESS

Contents

Contents

NSW Syllabus Outcomes

Units	1	2	3	4
NUMBER AND ALGEBRA				
MA2-RN-01 Representing numbers using place value				
Applies an understanding of place value and the role of zero to represent numbers to at least tens of thousands				
MA2-RN-02				
Represents and compares decimals up to 2 decimal places using place value				
MA2-AR-01 Additive relations				
Selects and uses mental and written strategies for addition and subtraction involving 2- and 3-digit numbers				
MA2-AR-02				
Completes number sentences involving addition and subtraction by finding missing values				
MA2-MR-01 Multiplicative relations				
Represents and uses the structure of multiplicative relations to 10 × 10 to solve problems				
MA2-MR-02				
Completes number sentences involving multiplication and division by finding missing values				
MA2-PF-01 Partitioned fractions				
Represents and compares halves, quarters, thirds and fifths as lengths on a number line and their related fractions formed by halving (eighths, sixths and tenths)				
MEASUREMENT AND SPACE				
MA2-GM-01 Geometric measure				
Uses grid maps and directional language to locate positions and follow routes				
MA2-GM-02				
Measures and estimates lengths in metres, centimetres and millimetres				
MA2-GM-03				
Identifies angles and classifies them by comparing to a right angle				
MA2-2DS-01 Two-dimensional spatial structure				
Compares two-dimensional shapes and describes their features				
MA2-2DS-02				
Performs transformations by combining and splitting two-dimensional shapes				
MA2-2DS-03				
Estimates, measures and compares areas using square centimetres and square metres				
MA2-3DS-01 Three-dimensional spatial structure				
Makes and sketches models and nets of three-dimensional objects including prisms and pyramids				
MA2-3DS-02				
Estimates, measures and compares capacities (internal volumes) using litres, millilitres and volumes using cubic centimetres				
MA2-NSM-01 Non-spatial measure				
Estimates, measures and compares the masses of objects using kilograms and grams				
MA2-NSM-02				
Represents and interprets analog and digital time in hours, minutes and seconds				
STATISTICS AND PROBABILITY				
MA2-DATA-01 Data				
Collects discrete data and constructs graphs using a given scale				
MA2-DATA-02				
Interprets data in tables, dot plots and column graphs				
MA2-CHAN-01 Chance				
Records and compares the results of chance experiments				
MAO-WM-01 Working mathematically				
Develops understanding and fluency in mathematics through exploring and connecting mathematical concepts, choosing and applying mathematical techniques to solve problems, and communicating their thinking and reasoning coherently and clearly				

		5	6	7	8	9	10	11	12	13	14	15	16	17	18	19	20	21	22	23	24	25	26	27	28	29	30	31	32	33	34	35
NUMBER AND ALGEBRA																																
MA2-RN-01	Representing numbers using place value																															
MA2-RN-02																																
MA2-AR-01	Additive relations																															
MA2-AR-02																																
MA2-MR-01	Multiplicative relations																															
MA2-MR-02																																
MA2-PF-01	Partitioned fractions																															
MEASUREMENT AND SPACE																																
MA2-GM-01	Geometric measure																															
MA2-GM-02																																
MA2-GM-03																																
MA2-2DS-01	Two-dimensional spatial structure																															
MA2-2DS-02																																
MA2-2DS-03																																
MA2-3DS-01	Three-dimensional spatial structure																															
MA2-3DS-02																																
MA2-NSM-01	Non-spatial measure																															
MA2-NSM-02																																
STATISTICS AND PROBABILITY																																
MA2-DATA-01	Data																															
MA3-DATA-02																																
MA2-CHAN-01	Chance																															
MAO-WM-01	Working mathematically																															

Number and Algebra

SET 1 Basic

1 6 + 2

2 6 + 4

3 2 + 2

4 5 + 4

5 7 + 1

6 8 + 3

7 9 + 10

8 2 + 6

9 8 − 3

10 7 − 1

11 3 − 3

12 10 − 2

13 2, 4, 6, ☐

14 1, 3, 5, ☐

15

I had a dozen eggs but I dropped 7. How many eggs were left?

☐ eggs

SET 2 Addition facts

1 Add 5 and 4.

2 Sum of 9 and 8

3 6 plus 5

4 What is the sum of 9 and 7?

5 What is 5 more than 13?

6 What is 6 more than 7?

7 What is 9 plus 7?

8 13 + 0

9 9 + ☐ = 15

10 Add $7 to $13.

11 11 cm + 5 cm = ☐ cm

12 11 plus 8

13 What number added to itself makes 18?

14 John spent $9 and $5. How much did he spend in total?

15

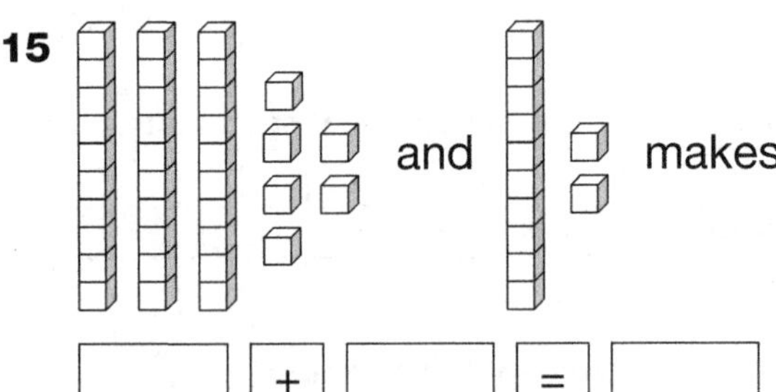

☐ + ☐ = ☐

Space Three-dimensional objects

Match the named 3D objects to the everyday objects.

1

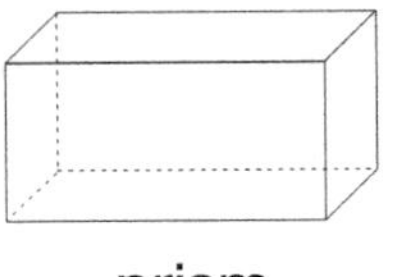

prism

2

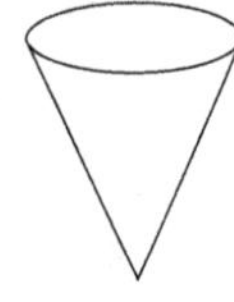

cone

3

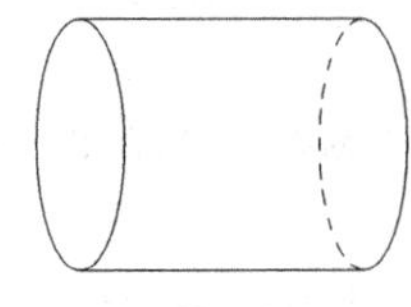

cylinder

Number and Algebra

SET 3 Skip counting

Complete the skip counting patterns.

1	0	3	6	9			
2	12	14	16	18			
3	40	50	60	70			
4	0	5	10	15			
5	6	10	14	18			
6	20	23	26	29			
7	30	35	40	45			

Complete each pattern and then write the rule.

8	50	55	60	65			

Rule: ____________

9	36	34	32	30			

Rule: ____________

10	21	26	31	36			

Rule: ____________

SET 4 Extension

1 55 = ☐ tens + ☐ ones

2 Which is smaller: 17 + 6 or 28 + 4?

3 $9 + $15

4 5 tens + 6 ones

5 Write 15 in words. ☐

6 $3 + $13 + $22

Shade two numbers in each circle that total 27.

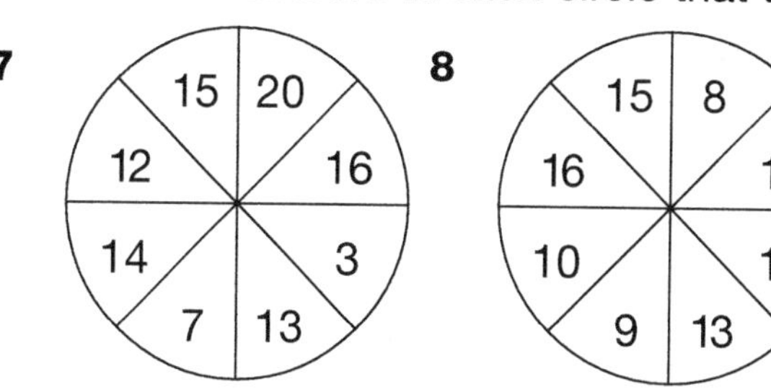

Working Mathematically

Find 4 pairs of numbers that have a total of 20.

2 3 16 15 17 5 4 18

9 ________ + ________ = 20

10 ________ + ________ = 20

11 ________ + ________ = 20

12 ________ + ________ = 20

Measurement Centimetres

Use the 1 centimetre dot paper to draw the following lines. The first one is done for you.

5 cm •—•—•—•—•—• • • • • • • • • • •

6 cm • • • • • • • • • • • • • • • •

12 cm • • • • • • • • • • • • • • • •

15 cm • • • • • • • • • • • • • • • •

13 cm • • • • • • • • • • • • • • • •

UNIT 2

Number and Algebra

SET 1 Basic

1 7 + 2

2 6 + 4

3 5 + 6

4 9 + 2

5 9 + 4

6 5 + 6

7 3 + 7

8 10 – 5

9 7 – 4

10 9 – 6

11 12 – 3

12 Half of 10

13 Double 6.

14 7 + ☐ = 10

15

I had 11 marbles and I found 8 more. How many do I have now?

☐ marbles

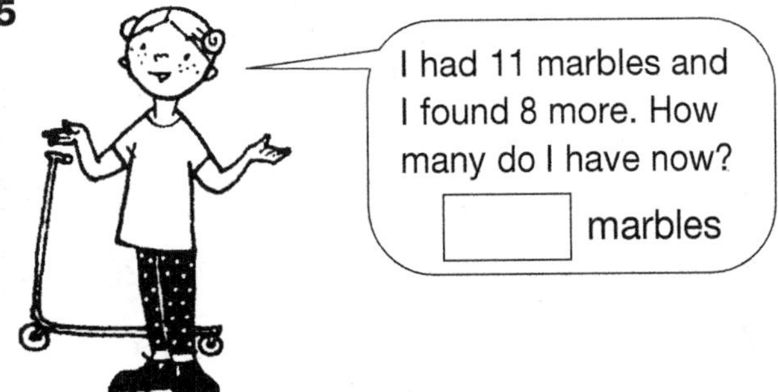

SET 2 Subtraction facts

1 Take 10 from 20.

2 Subtract 5 from 17.

3 Subtract 8 from 15.

4 Take $4 from $13.

5 What is the difference between $7 and $19?

6 What is the difference between 7 and 4?

7 13 km less than 20 km

8 Take 14 cm from 64 cm.

9 Subtract 6 ones from 2 tens.

10 15 less than 19

11 Take 11 minutes from 20 minutes.

12 Subtract 7 from 13.

13 By how much is 13 less than 18?

14 Deduct 1 from 12.

15 Deduct 13 pages from 17 pages.

16 Subtract 4 from 20.

Space Symmetry

Complete each shape by using the line of symmetry as the starting point.

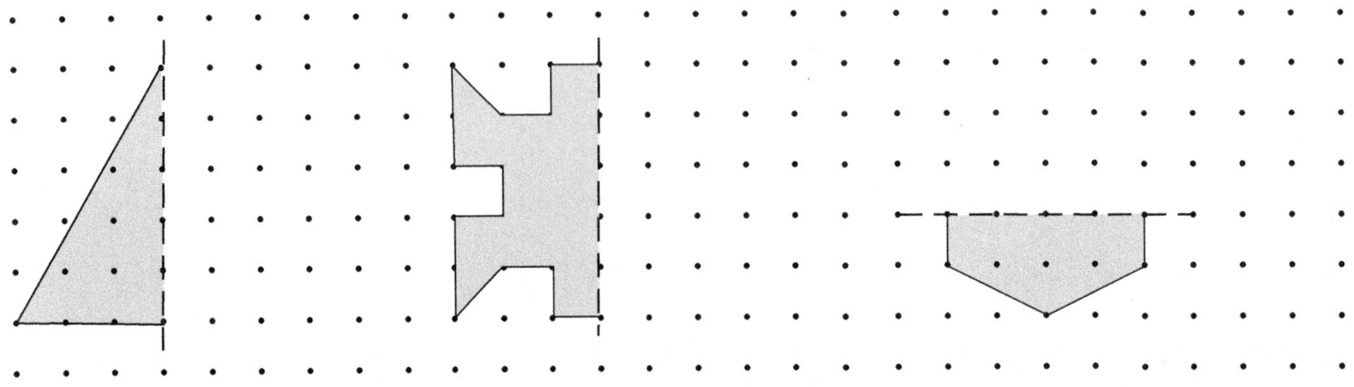

Number and Algebra

SET 3 Numbers to 1000

1 Write 123 in words.

2 Write 367 in words.

3 Write the number two hundred and thirty-seven.

4 Write the number four hundred and sixty.

5 Write the missing numbers.

390, 380, ___, ___

6 How many hundreds in 515?

7 What is the absent number?

501, 401, ___, 201

8 Write the number 10 more than 6 tens.

9 How many tens in 100?

Draw beads on the abacuses to represent the numbers.

10

Hund Tens Ones

271

11

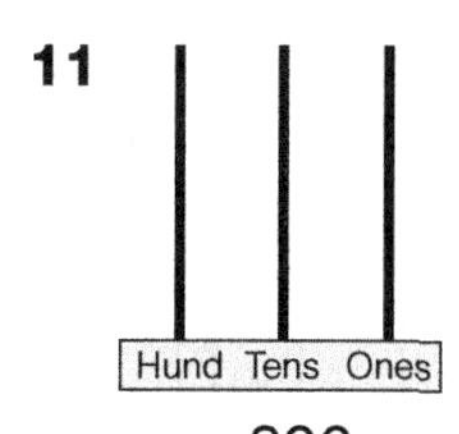

806

SET 4 Extension

1 16 – 12

2 48 – 6

3 15 minus 6

4 Eighteen take away 6

5 What season comes after spring?

6 How many days in May?

7 When is Australia Day?

8 If Monday was 3 June, what would be the date of the following Wednesday?

9 How many 10c coins in $2?

10 What number is between 66 and 68?

11 Share 12 lollies among 4.

12 15 + 6 – 2

13 14 km minus 4 km

14 13 + 8 – 1

15 Write 3 subtraction number sentences that have an answer of 12.

___ – ___

___ – ___ = 12

___ – ___

Measurement Area

Working Mathematically

Can you create 4 different shapes that have an area of 5 squares?
Colour each one a different colour.

Number and Algebra

SET 1 Basic

1 6 + 5

2 4 + 3

3 9 + 2

4 5 + 4

5 6 + 0

6 8 + 5

7 9 – 5

8 8 – 7

9 10 – 10

10 10 – 4

11 7 take away 3

12 10 take away 9

13 Half of 12

14 Double 7.

Jimi ate 15 peanuts and Lauren ate 7. How many peanuts did they eat altogether?

☐ peanuts

SET 2 Changing addends to make 10

By looking for numbers that add up to make 10, the addition sum becomes easier for you to solve.

Look for pairs of numbers that make 10.

1 8 plus 3 plus 7

2 18 plus 3 plus 17

3 5 plus 8 plus 5

4 Add 6 and 16 and 4.

5 Add 23 and 6 and 7.

6 Add 16 and 6 and 4 and 4.

7 The sum of 11 and 5 and 9

8 The sum of 17 and 5 and 13

9 The sum of 8 and 5 and 12

10 3 + 16 + 7 + 4

11 2 plus 4 plus 8 plus 6

12 14 plus 7 plus 6 plus 3

13 Add 16 and 8 and 4.

14 18 + ☐ = 25

Space Describing position

Petri built a model out of blocks. Put the letters on the blocks to discover the secret message.

1 Put a **C** on the top left block.

2 Put an **N** on the middle block.

3 Put a **D** on the bottom left block.

4 Put a **T** on the top right block.

5 Put a **G** on the bottom right block.

6 Put another **D** to the right of **N**.

7 Put an **A** to the left of **N**.

8 Put an **O** between **D** and **G**.

9 Put another **A** between **C** and **T**.

10 What is the message?

____

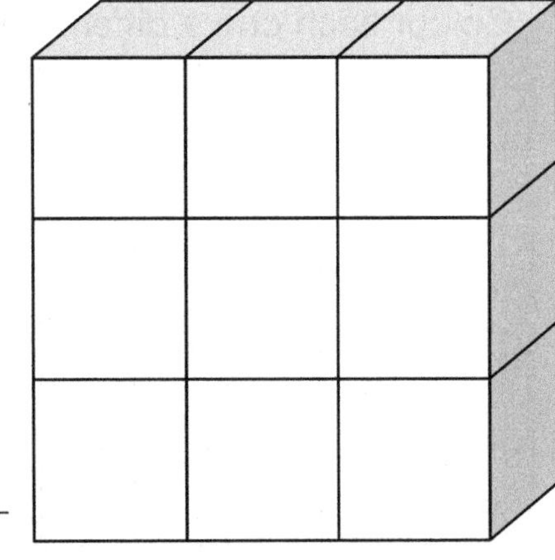

Number and Algebra

SET 3 Equal groups/multiplication

Write number sentences to describe the groups.

1 ▢ ▢ / ▢ ▢

2 groups of □ = □

2 ▢▢ ▢▢ / ▢▢ ▢▢

□ groups of □ = □

3 ○○ ○○ ○○ / ○○ ○○ ○○

□ groups of □ = □

4 ○○○ ○○○ ○○○ ○○○ / ○○ ○○ ○○ ○○

□ groups of □ = □

5 ○○○ ○○○ ○○○ ○○○ / ○○○ ○○○ ○○○ ○○○

□ groups of □ = □

SET 4 Extension

1 69 = □ tens + □ ones

2 Which is larger: 17 + 9 or 8 + 10?

3 Forty-five and six more

4 49 + 18

5 How many eyes would 8 children have?

6 19 = □ ten and □ ones

7 Which is larger, 8 tens and 9 ones or 9 tens?

8 25 plus 29

9 What number is between 40 and 42?

10 Which season comes after summer?

11 38 + 9

12 36 = □ tens + □ ones

13 50, 55, □, 65, □

14 3 tens + 8 ones

Working Mathematically

Shade two numbers in each circle that multiply to give 24.

15

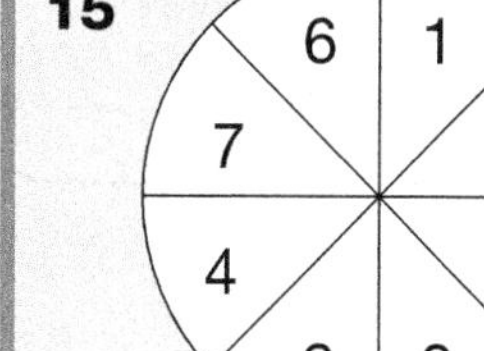

16

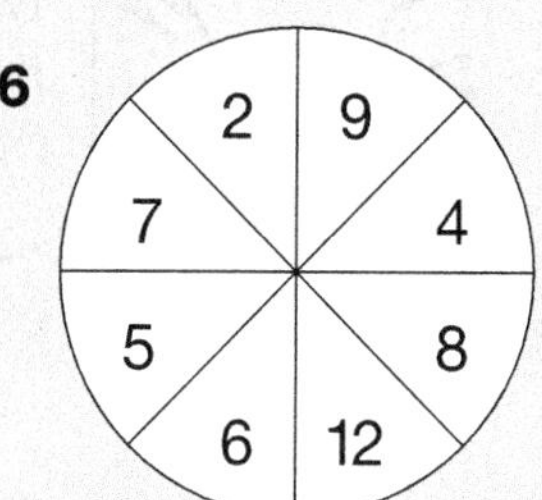

Statistics and Probability Tally marks

Teachers' cars

Toyota	|||
Holden	𝍸 𝍸 |||
Ford	𝍸 𝍸
Mazda	|||
Nissan	||
Mitsubishi	𝍸 |

1 How many teachers drive Fords?

2 How many drive Holdens?

3 How many drive Toyotas?

4 Which is the most popular car at this school?

□

5 Which cars have equal popularity?

□

UNIT 4

Number and Algebra

SET 1 Basic

1 10 + 4

2 7 + 6

3 7 + 7

4 8 + 5

5 2 + 12

6 13 + 4

7 8 + 7

8 9 + 5

9 11 – 2

10 13 – 2

11 14 – 2

12 16 – ☐ = 14

13 Double 8.

14 Half of 16

15

Susan had $6 but Peter had twice as much. How much money did Peter have?

$ ☐

SET 2 Inverse operations

Check these addition facts using subtraction. Put a tick (✓) for correct facts and a cross (✗) for incorrect ones.

1 5 + 8 = 13 13 – 8 = ☐ ☐

2 12 + 9 = 20 20 – 9 = ☐ ☐

3 7 + 11 = 18 18 – 11 = ☐ ☐

4 8 + 6 = 14 14 – 6 = ☐ ☐

5 13 + 6 = 18 18 – 6 = ☐ ☐

Solve these subtraction problems, then check your answers using addition.

6 A golfer had 12 balls and lost 7 in the scrub. How many balls are left?

☐ – ☐ = ☐ ☐ + ☐ = ☐

7 Jacqui picked 18 strawberries. If 5 were rotten, how many could she eat?

☐ – ☐ = ☐ ☐ + ☐ = ☐

Space Face, edge and vertex

Use the words face, edge and vertex to label the prisms.

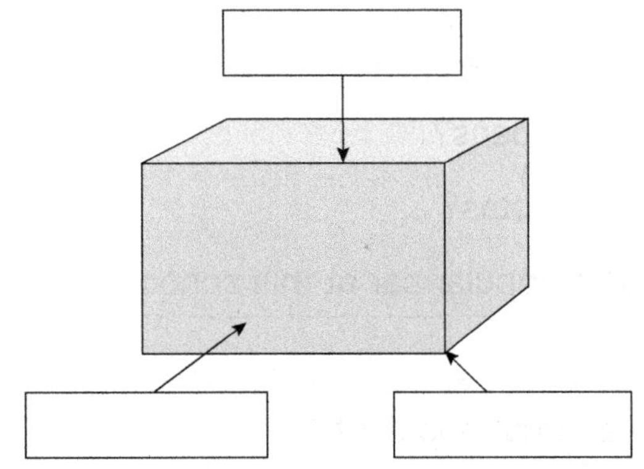

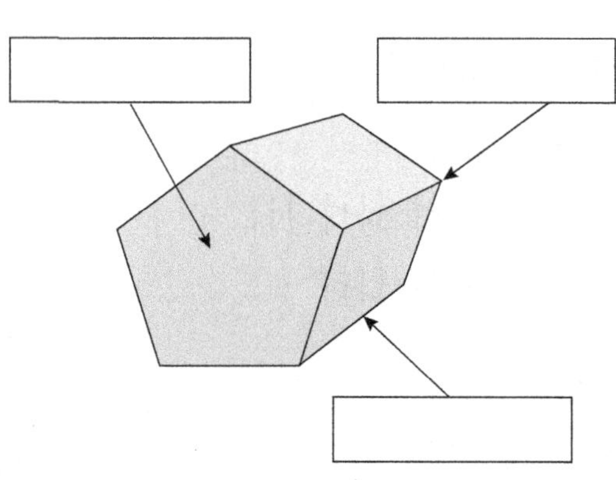

Number and Algebra

SET 3 Multiplication facts (2s)

1 Use the array to help you answer the questions.

● ●	1	1 × 2 = ______
● ●	2	2 × 2 = ______
● ●	3	3 × 2 = ______
● ●	4	4 × 2 = ______
● ●	5	5 × 2 = ______
● ●	6	6 × 2 = ______
● ●	7	7 × 2 = ______
● ●	8	8 × 2 = ______
● ●	9	9 × 2 = ______
● ●	10	10 × 2 = ______

Working Mathematically

2 How many different arrays can you make with 20 buttons? ______
Draw one of your arrays.

SET 4 Extension

1 What number is 3 more than 72?

2 What number is 15 less than 20?

3 What is the next odd number after 9?

4 How many days in 2 weeks?

5 Which is the larger number: 132 or 123?

6 How much change would I get from $1 if I spent 80c?

7 Share 20 counters among 4.

8 Ten more than seventeen

9 81 = ☐ tens + ☐ ones

10 How many 5c coins make 50c?

11 30 + 50

12 Which number is larger, 176 or 167?

13 Write 92 in words.

☐

14 How many eggs in 2 dozen?

15 What month comes before June?

16 Name this object.

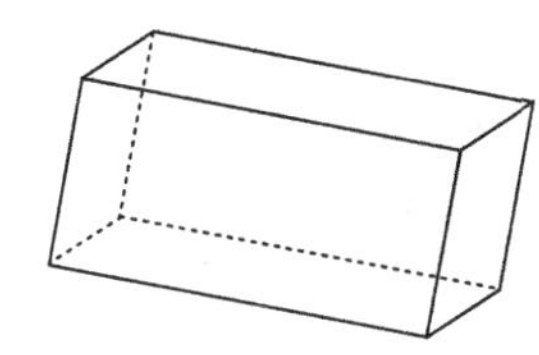

Measurement Measuring in centimetres

Estimate and then measure the length of these pencils in centimetres.

1 Estimate ☐ Actual ☐

2 Estimate ☐ Actual ☐

3 Estimate ☐ Actual ☐

4 Estimate ☐ Actual ☐

5 Estimate ☐ Actual ☐

Number and Algebra

SET 1 Basic

1 4 + 7

2 3 + 9

3 $7 + $5

4 8 – ☐ = 7

5 ☐ – 9 = 1

6 3 × 4

7 7 × 2

8 6 × 4

9 ☐ – 5 = 2

10 17 + 3

11 Half of 40

12 Double 20.

13 2 lots of 4

14 5, 10, 15, ☐

15

SET 2 Odd and even numbers

Model these numbers on the dot paper.
Then tick whether they are odd or even.
The first one is done for you.

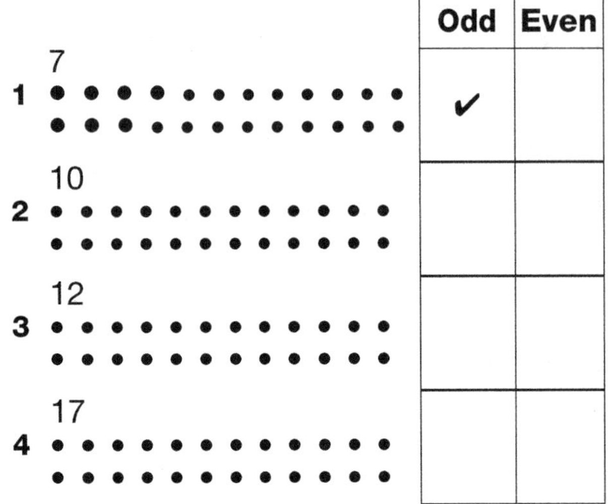

		Odd	Even
1	7	✓	
2	10		
3	12		
4	17		

Write odd or even after the following numbers.

5 23 ____________

6 44 ____________

7 63 ____________

8 88 ____________

9 96 ____________

10 123 ____________

11 134 ____________

12 139 ____________

13 486 ____________

Space Triangles

Draw a line to match the descriptions with the triangles.

I am a triangle, all my sides are the same length and all of my angles are the same size. 

I am a triangle, 2 of my sides are the same length and 2 of my angles are the same size. 

I am a triangle, none of my sides are the same length and none of my angles are the same size. 

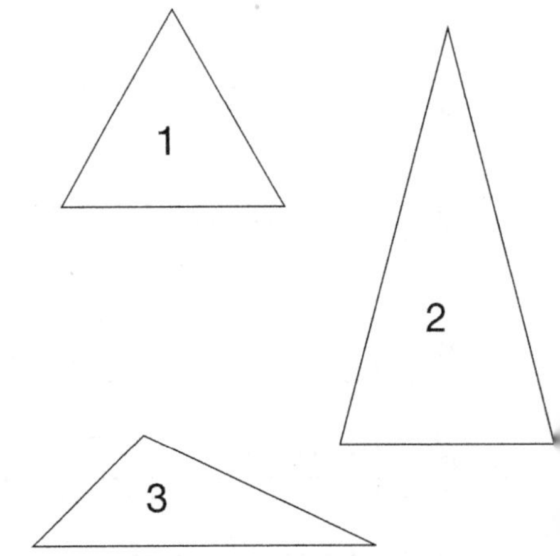

Number and Algebra

SET 3 Connecting addition and subtraction

Complete these addition facts. Then use the subtraction facts to check your answers.

1	6 + 7 = ☐	13 − 6 = ☐
2	8 + 6 = ☐	14 − 8 = ☐
3	10 + 9 = ☐	19 − 10 = ☐
4	12 + 5 = ☐	17 − 12 = ☐
5	3 + 13 = ☐	16 − 3 = ☐
6	18 + 2 = ☐	20 − 18 = ☐
7	9 + 9 = ☐	18 − 9 = ☐

8 Lisa cooked 18 biscuits. If she burnt 7, how many were left?

9 Henry bought a dozen apples and gave 5 away. How many apples were left?

10 James saved $36 but spent $8. How much did he have left?

SET 4 Extension

1 15 + ☐ = 20

2 93 = ☐ tens and ☐ ones

3 How many rows of 5 make 20?

4 If one pencil costs 10c, how much would 4 pencils cost?

5 Write 12 in words. ☐

6 How many days in April?

7 How many months in a season?

8 What month comes after August?

9 How many eggs in $1\frac{1}{2}$ dozen?

10 Estimate an answer to 19 + 21.

11 Write the number that is 6 tens and 7 ones.

12 Jill carried 13 books but dropped 5.

She had ☐ books left.

13 Is 127 an even number?

14 Add 27 and 18.

15 How many hours from 3 pm to 9 pm?

16 Name this shape.

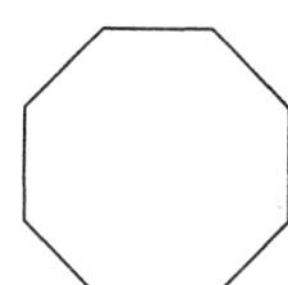

Measurement Time in minutes

Draw the minute hands on the clock faces.

1 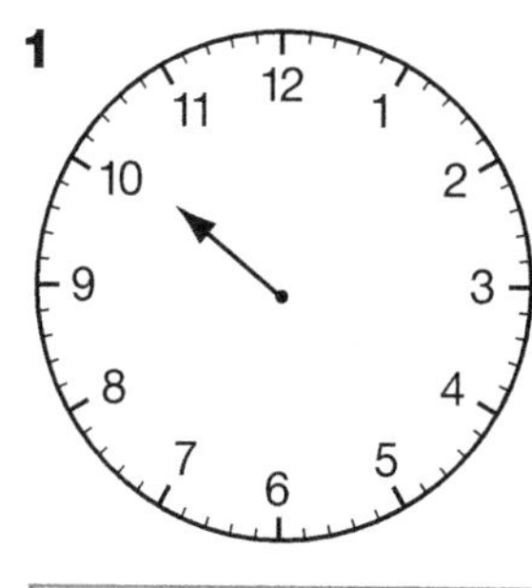

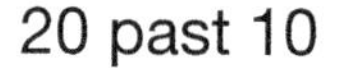

20 past 10

2 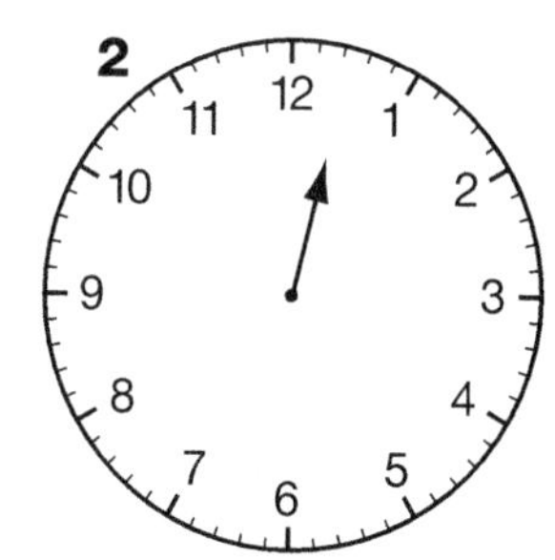

25 past 12

3 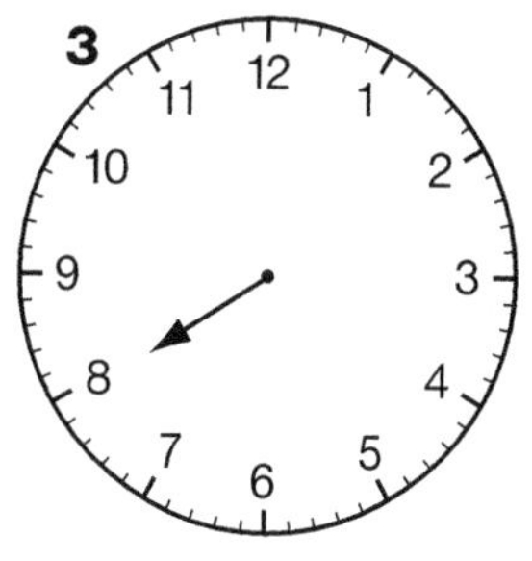

5 past 8

4 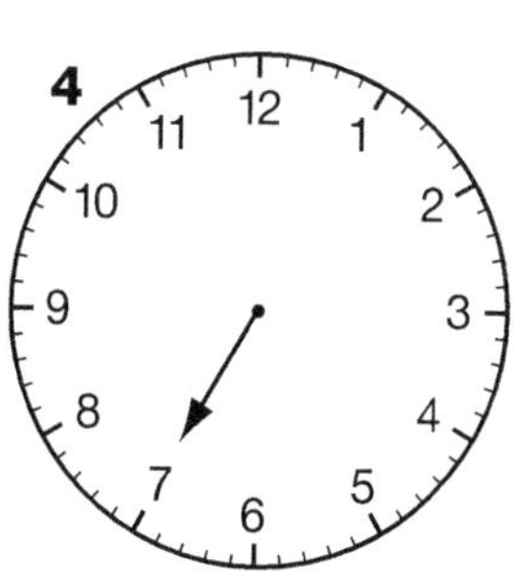

10 past 7

UNIT 6

Number and Algebra

SET 1 Basic

1 7 – 6

2 9 – 5

3 9 – 0

4 1 + 6

5 4 + 5

6 0 + 9

7 2 + 10

8 Double 2.

9 10 – 4

10 9 – 6

11 One less than 15

12 Two more than 9

13 20c – 5c

14 10 – 10

15

I saved $6 in March and $14 in April. How much have I saved so far?

$ ☐

SET 2 Expanding and comparing numbers

Expand these numbers.

1 539 = 500 + ☐ + ☐

2 635 = ☐ + ☐ + ☐

3 278 = ☐ + ☐ + ☐

4 386 = ☐ + ☐ + ☐

5 794 = ☐ + ☐ + ☐

6 497 = ☐ + ☐ + ☐

Working Mathematically

Use the $<$ or $>$ symbols to complete these sentences.

The first one is done for you.

7 400 + 20 + 8 $<$ 600

8 600 + 80 + 5 ☐ 700

9 700 + 50 + 2 ☐ 700

10 300 + 90 + 4 ☐ 500

11 500 + 70 + 6 ☐ 500

Space North, south, east and west

1 Which block is north of O? ☐

2 Which block is south of D? ☐

3 Which block is south of C? ☐

4 Which block is north of J? ☐

5 Which block is east of E? ☐

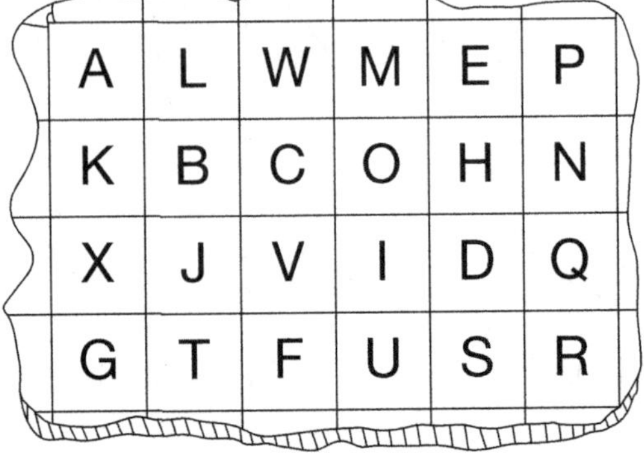

Number and Algebra

SET 3 Multiplication facts (5s)

1 [2] × 5 = 10

2 4 × [ ] = 20

3 [ ] × 5 = 20

4 [ ] × 5 = 30

5 [ ] × 5 = 15

6 7 × [ ] = 35

7 5 × 5 = [ ]

8 8 × 5 = [ ]

9 10 lots of 5 = [ ]

10 9 × 5 = [ ]

11 A farmer has 6 paddocks with 3 sheep in each. How many sheep does he have altogether?

12 Peta earns $8 every hour she works. How much does she earn in 3 hours?

13 Rana saves $7 each week. How much would she save in 3 weeks?

SET 4 Extension

1 15 – 13

2 36 – 5

3 14 minus 8

4 Nine take away 4

5 15c – 10c

6 What season comes after winter?

7 How many days in April plus September?

8 Subtract 10 from two dozen.

9 Forty cents less two 10c coins

10 What number is halfway between 70 and 80?

11 4 + 41 – 2

12 150 + 30

13 Estimate an answer to 28 + 31.

14 Share 20 among 5.

Complete the grid.

	+	12	14	16	18	20	22
15	6						
16	12						

Statistics and Probability Dot plots

Use the tallied information to complete the dot plot.

Favourite drinks

Cola	𝍸 |||
Orange juice	𝍸 𝍸	
Lemonade	||||
Lime cordial		
Creaming soda	𝍸

Favourite drinks

Cola | Orange juice | Lemonade | Lime cordial | Creaming soda

Drinks

UNIT 7

Number and Algebra

SET 1 Basic

1 7 + 2

2 5 + 4

3 2 + 3

4 5 + 5

5 8 + 2

6 8 – 4

7 7 – 4

8 9 – 6

9 Half of 12

10 Double 7.

11 2, ☐, ☐, 8, 10, ☐

12 10 + 4

13 7 + 7

14 11 – 2

15

Blake had 18 marbles but gave half away. How many marbles were left?

☐ marbles

SET 2 The division symbol

1 How many groups of 4 above?

2 How many groups of 3 above?

3 How many groups of 2 above?

4 How many groups of 6 above?

Use the dogs to solve the divisions.

5 16 ÷ 2 = ☐

6 16 ÷ 4 = ☐

7 16 ÷ 1 = ☐

8 16 ÷ 8 = ☐

9 16 ÷ 16 = ☐

Space Parallel lines

Use the dots to draw lines parallel to the ones given.

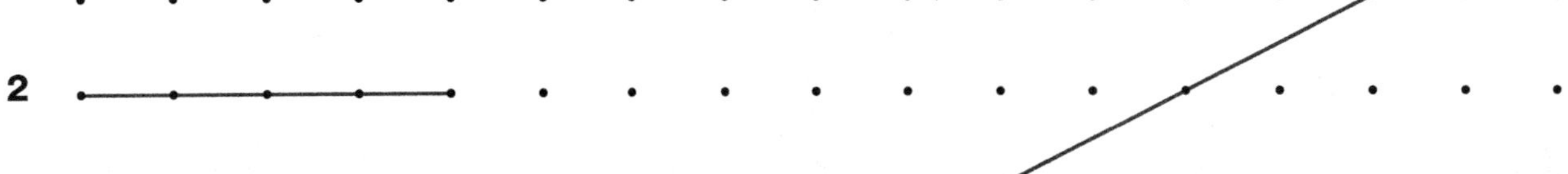

Number and Algebra

SET 3 Addition

1 Add 15 and 4.

2 10 + 15

3 26 plus 5

4 What is the sum of 17 and 7?

5 What is 5 more than 33?

6 What is 6 more than 18?

7 What is 39 plus 7?

8 93 + 0

9 49 + ☐ = 55

10 Add $7 to $13.

11 11 cm + 20 cm = ☐ cm

12 77 plus 8

Use the number lines to solve these.

13 22 + 13 =

20 21 22 23 24 25 26 27 28 29 30 31 32 33 34 35 36 37 38 39 40

14 29 + 11 =

20 21 22 23 24 25 26 27 28 29 30 31 32 33 34 35 36 37 38 39 40 41

SET 4 Extension

1 Share $15 among 3.

2 What is the next odd number after 79?

3 Which is the larger number, 64 or 46?

4 How many 5c coins make 80c?

5 Write 16 in words. ☐

6 Is 39 an even number?

7 What number comes 3 before 72?

8 How many girls are there if there are 6 rows of 4 girls?

9 How many 20c coins in $1.80?

10 What is the fifth month of the year?

11 Place the numbers 2, 4 and 7 in the circles so that the sum of each line is 15.

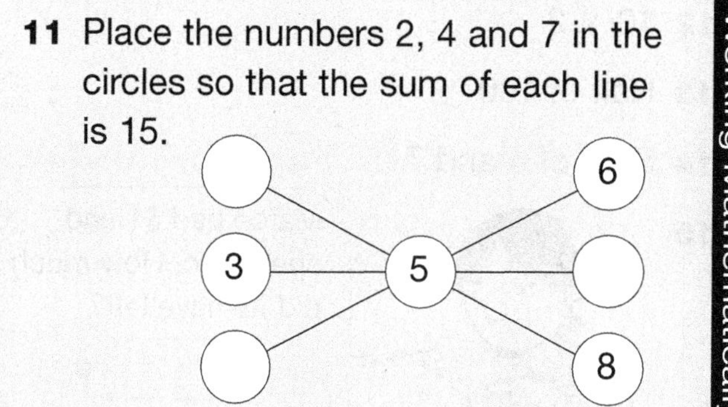

Working Mathematically

Measurement Informal capacity

Circle the most suitable container in each set to measure the capacity of the bathtub and the saucepan.

1

2

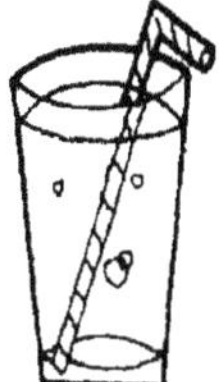

Number and Algebra

SET 1 Basic

1 $3 + 3$

2 $4 + 4$

3 $8 + 3$

4 $6 + 3$

5 $19 + \square = 20$

6 $9 - 5$

7 $9 - 3$

8 $9 - 2$

9 $9 - 9$

10 7×4

11 8×4

12 10×2

13 Half of 100

14 Sum of 4 and 7

15

Marco had $1 and spent 50c. How much did he have left? $\square$ c

SET 2 Subtraction patterns

1 $7 - 2$

2 $70 - 20$

3 $9 - 5$

4 $90 - 50$

5 $13 - 6$

6 $130 - 60$

7 $8 - 4$

8 $80 - 40$

9 $800 - 400$

10 $500 - 300$

Solve the problems.

11 Barry's mass is 70 kg and Jin's mass is 40 kg. What is the difference in their masses?

12 Hannah needs $800 to buy a new washing machine. If she has saved $300, how much more does she need to save?

Space Pyramids

1 Shade all pyramids.

2 Tick the pyramid whose base is an octagon.

3 Put a cross on the pyramid with 6 corners.

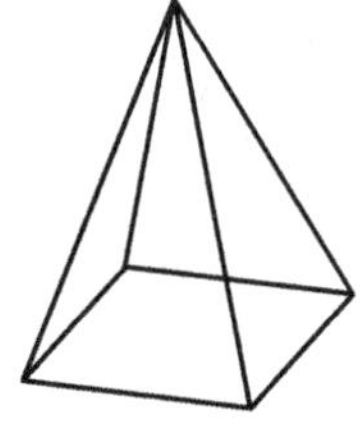

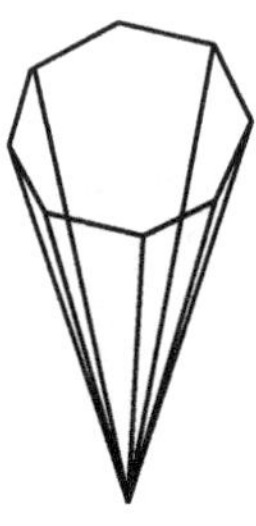

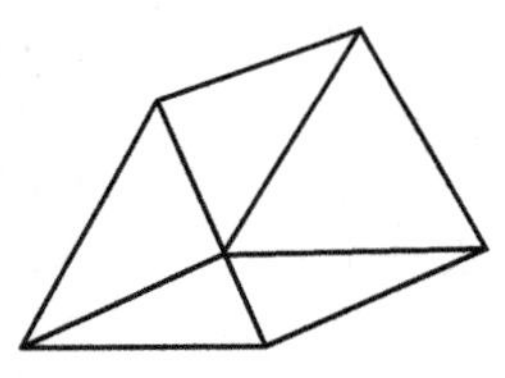

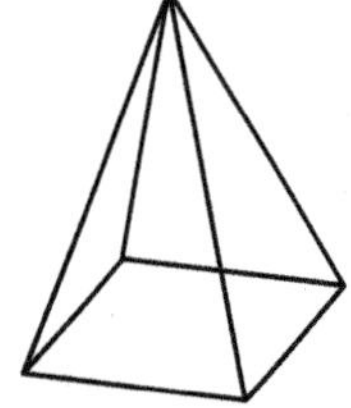

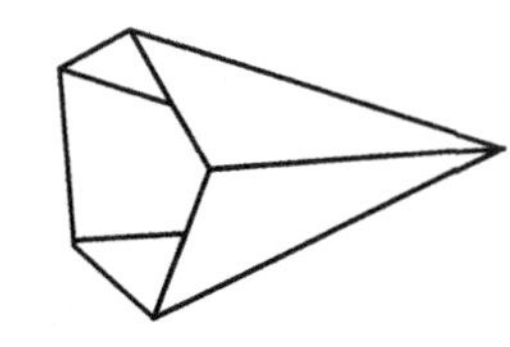

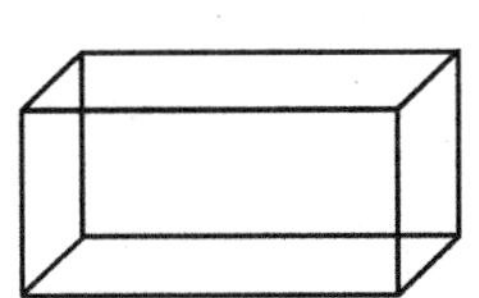

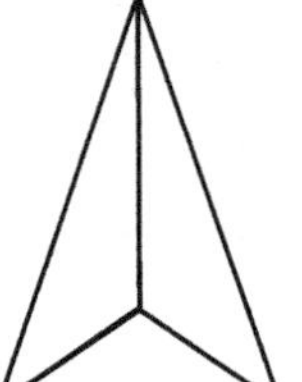

Number and Algebra

SET 3 Halves, quarters and eighths

Colour the fraction of each circle.

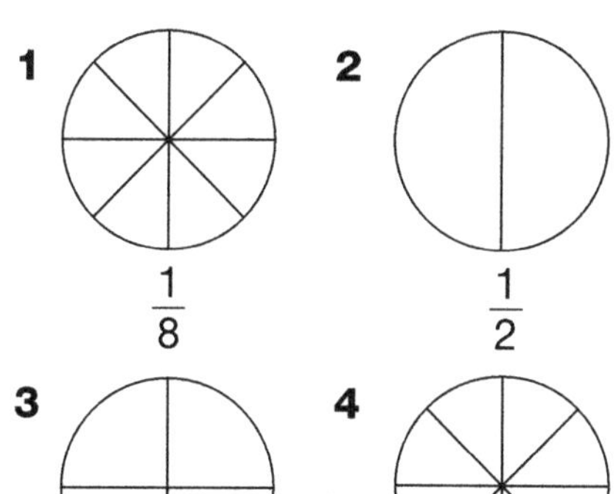

1 $\frac{1}{8}$ 2 $\frac{1}{2}$

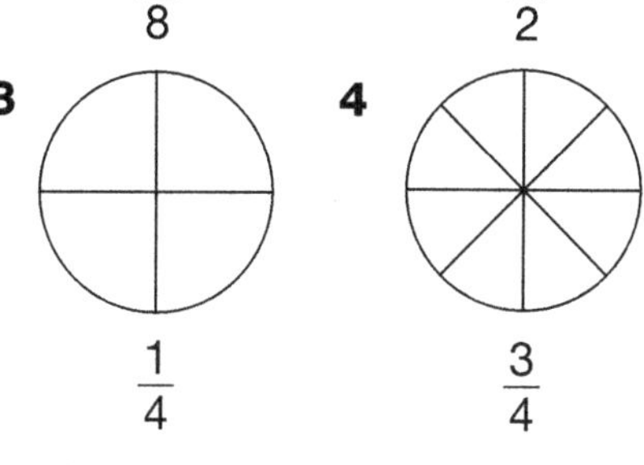

3 $\frac{1}{4}$ 4 $\frac{3}{4}$

5 Order the fractions from smallest to largest by numbering them 1 to 4.

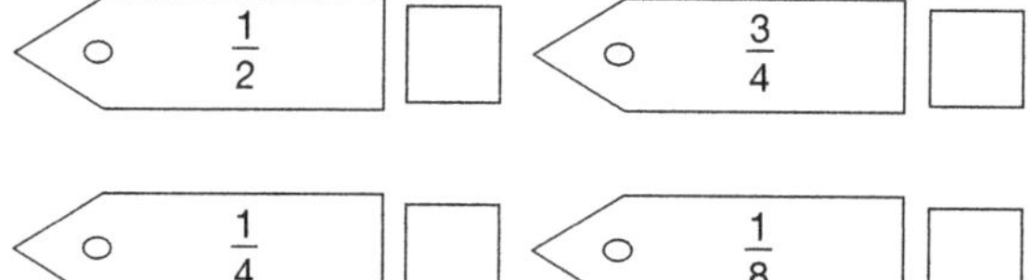

$\frac{1}{2}$ $\frac{3}{4}$ $\frac{1}{4}$ $\frac{1}{8}$

6 Circle the groups into the given fractions.

a $\frac{1}{2}$ b $\frac{1}{4}$

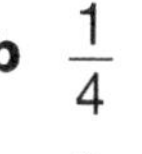

SET 4 Extension

1 4 + 4 + 4

2 15 – 0

3 Share 18 among 3.

4 Which is the third month of the year?

5 130 + 60

6 Write 17 in words.

7 How many wheels are there on 7 push bikes?

8 Subtract 5 from 91.

9 What is the sum of 17 and 24?

10 What is the sum of 19 and 12?

11 How many days are in June?

12 How many halves are in a whole?

13 Estimate an answer to 42 + 23.

14 Write 4 number sentences that equal 27.

☐ + ☐
☐ + ☐
☐ – ☐ = 27
☐ – ☐

Working Mathematically

Measurement Calculating area

Tick the shape with an area of 12 squares. Put a cross on the shape with an area of 14 squares.

Number and Algebra

SET 1 Basic

1 8 – 5

2 10 – 4

3 10 – 1

4 2 + 7

5 3 + 6

6 1 + 8

7 3 + 9

8 Double 9

9 10c – 5c

10 8 – 7

11 Two less than 17

12 Two more than 8

13 6 × 2

14 6 × 4

15

John has 6 popsticks in one hand and 9 in his other hand. How many popsticks is John holding altogether?

☐ popsticks

SET 2 Split strategy for addition

Add the tens then the ones.
24 + 38 becomes 50 + 12 = 62.

Add the following numbers by adding the tens part then the ones part.

1 27 + 32 becomes ☐ + ☐ = ☐

2 44 + 23 becomes ☐ + ☐ = ☐

3 35 + 17 becomes ☐ + ☐ = ☐

4 58 + 22 becomes ☐ + ☐ = ☐

Complete the grid to show the sum of each pair of numbers.

	Numbers	Sum
5	29 and 16	45
6	10 and 23	
7	42 and 17	
8	16 and 32	
9	36 and 15	
10	45 and 23	

Statistics and Probability Picture graphs

Mr Black's class hair colours

Blond

Red

Brown

Green

Black

Fair

Use the graph to answer the questions.

1 How many children have blond hair? ☐

2 How many children have brown hair? ☐

3 How many children have red hair? ☐

4 How many children are in Mr Black's class? ☐

5 Which colours have the same number of children?

____

6 Write a question for this graph.

Number and Algebra

SET 3 Related multiplication facts

1 2 × 2

2 4 × ☐ = 8

3 3 × 4

4 6 × ☐ = 24

5 2 rows of 10

6 10 rows of 4

7 4 × ☐ = 32

8 How many sides on 4 squares?

9 How many legs on 10 birds?

10 How many pencils in 8 boxes of 5 pencils?

Working Mathematically

11 This array was made to show 5 × 8.
Write 4 other multiplication or division facts you can see inside the array.

◆◆◆◆◆◆◆◆	3	×	8	=	24
◆◆◆◆◆◆◆◆				=	
◆◆◆◆◆◆◆◆				=	
◆◆◆◆◆◆◆◆				=	
◆◆◆◆◆◆◆◆				=	

SET 4 Extension

1 4 × ☐ = 8

2 How many trees are there if there are 7 rows of 4 trees?

3 How many books are there if there are 10 bundles of 8 books?

4 What number is 7 more than 23?

5 Write 23 in words.

6 Share 16 among 4.

7 Which month has the least days?

8 How many sides on 6 rectangles?

9 What number comes 3 before 98?

10 Share 25 among 5.

11 85 = ☐ tens and ☐ ones

12 How many halves in 2 wholes?

13 What is the eighth month of the year?

14 10 pairs of socks = ☐ socks

15 Shade 4 numbers that add to make 47.

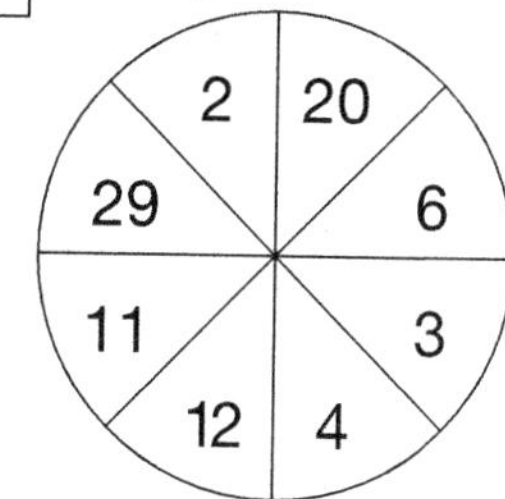

Measurement Digital times

Write each time in digital form.

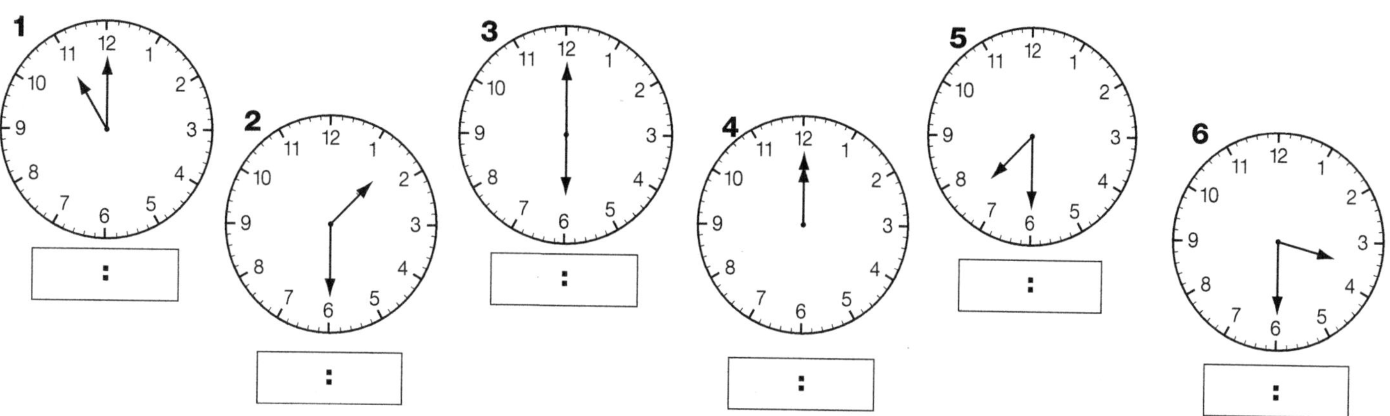

UNIT 10

Number and Algebra

SET 1 Basic

1 9 + 4

2 \$9 – \$7

3 3 × 4

4 10 + 2

5 12 – 2

6 6 × 4

7 4 × 4

8 9 take away 4

9 4 less than 17

10 Double 8.

11 90c + 5c

12 16 minus 10

13 20 plus 10

14 2 × 10

15 Trent saved \$6 one week and \$8 the next week. What were his total savings for the 2 weeks?

\$ ____

SET 2 Use addition to solve subtraction

Solve each subtraction fact then write an addition fact to check your answer.

	Subtraction	Addition
1	17 – 6 = 11	6 + 11 = 17
2	23 – 9 =	+ =
3	35 – 31 =	+ =
4	57 – 52 =	+ =
5	27 – 18 =	+ =
6	45 – 32 =	+ =

7

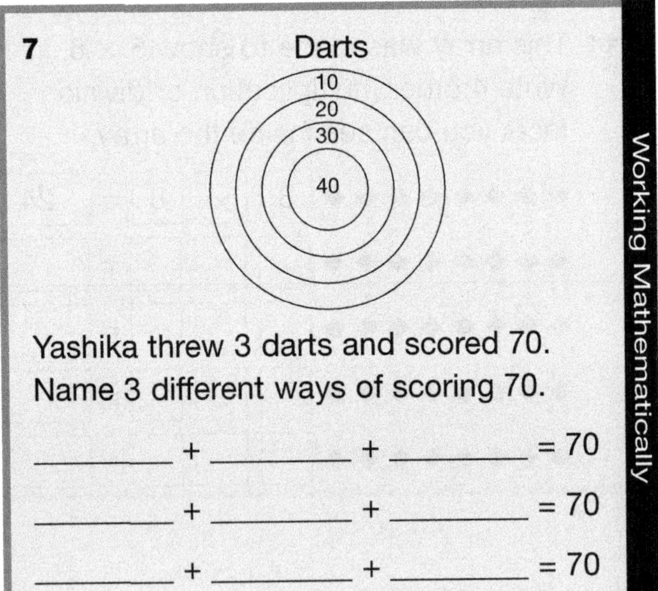

Yashika threw 3 darts and scored 70. Name 3 different ways of scoring 70.

____ + ____ + ____ = 70

____ + ____ + ____ = 70

____ + ____ + ____ = 70

Working Mathematically

Space Angles

Colour all the angles that are smaller than a right angle blue. Colour all the angles that are larger than a right angle red.

1

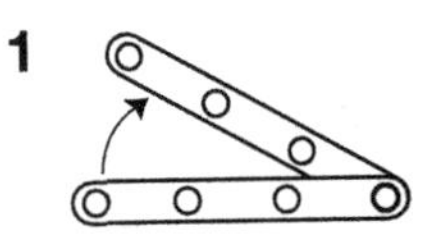

2

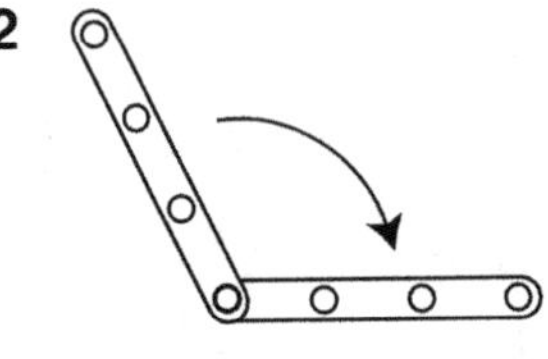

3

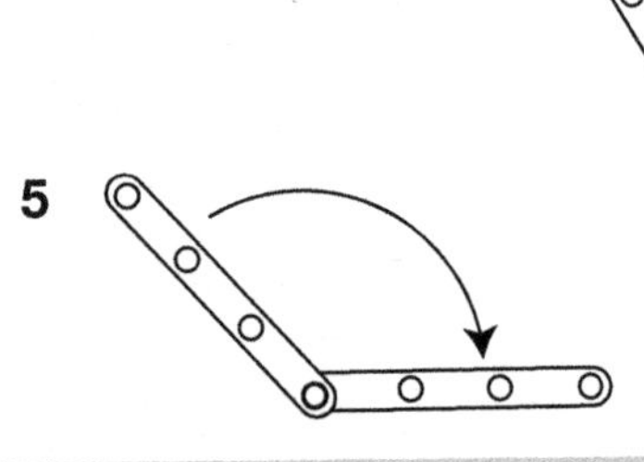

4

5

Number and Algebra

SET 3 Comparing and ordering fractions

Complete the sentences using the following signs.

= < >

1 $\frac{2}{8}$ ☐ $\frac{5}{8}$

2 $\frac{3}{8}$ ☐ $\frac{2}{8}$

3 $\frac{6}{8}$ ☐ $\frac{5}{8}$

4 $\frac{1}{2}$ ☐ $\frac{4}{8}$

5 $\frac{4}{8}$ ☐ $\frac{6}{8}$

6 $\frac{7}{8}$ ☐ $\frac{3}{8}$

7 $\frac{1}{2}$ ☐ $\frac{2}{2}$

8 $\frac{4}{4}$ ☐ $\frac{2}{2}$

9 $\frac{8}{8}$ ☐ $\frac{3}{4}$

10 $\frac{1}{2}$ ☐ $\frac{3}{4}$

Shade the shapes to match the fractions.

11 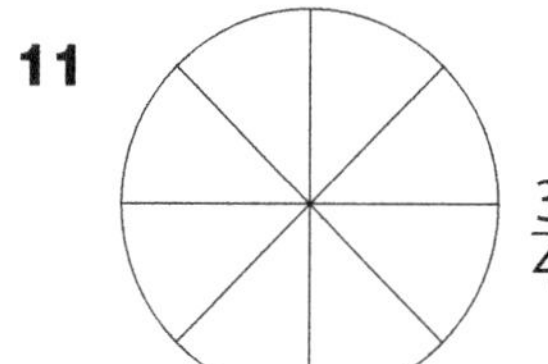 $\frac{3}{4}$

12 $\frac{1}{2}$

SET 4 Extension

1 82 = ☐ tens + ☐ ones

2 Which is larger: 6 + 4 or 3 + 8?

3 Twenty-one and nine more

4 35c + 55c

5 How many sides on 6 triangles?

6 If Wednesday is 1 February, what is Thursday's date?

7 10 + 10 + 10 + 5

8 How many $5 notes in $50?

9 Share 18 among 3.

10 Write 95 in words. ☐

11 Forty-six and sixteen more

12 234 = ☐ hund + ☐ tens + ☐ ones

13 35 – 17

14 Write 28 in words. ☐

15 Mt Cook Primary School has 37 infant children and 56 primary. How many children attend the school altogether?

Measurement Metres

Colour the correct boxes to record the heights of the following items.

	Item	Taller than 1 m	About the same as 1 m	Shorter than 1 m
1	Broom			
2	Chair			
3	Refrigerator			
4	Rubbish bin			
5	Shovel			
6	Telephone booth			

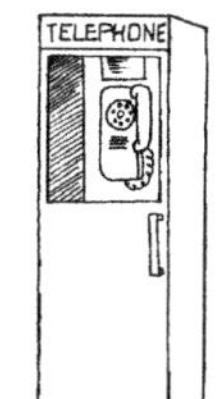

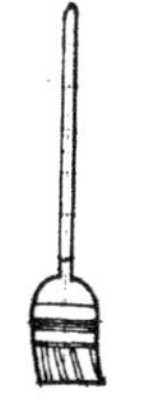

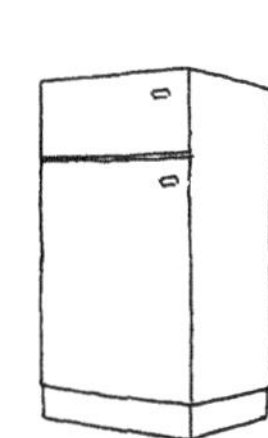

UNIT 11

Number and Algebra

SET 1 Basic

1 5 + 5

2 6 + 4

3 14 + 6

4 19 + 1

5 10 – 7

6 10 – 6

7 10 – 5

8 10 – 4

9 4 × 3

10 4 × 6

11 4 × 4

12 Sum of 11 and 9

13 6 plus 8

14 Six more than 9

15

Jill had $17 but Joyo had only $5. How much more money did Jill have?

$ ______

SET 2 Counting by tens

1 Count forward by tens.

14	24						

2 Count forward by tens.

130	140						

3 Count forward by tens.

57	67						

4 Count backward by tens.

93	83						

5 Count backward by tens.

100	90						

6 Count backward by tens.

255	245						

7 34, 44, 54, ____, ____

8 122, 132, 142, ____, ____

9 54, 64, 74, ____, ____

10 125, 135, 145, ____, ____

11 150, 160, 170, ____, ____

Number and Algebra Number patterns

Complete each pattern then write the rule for it.

1

4	6	8	10				

Rule: ______

2

3	6	9	12				

Rule: ______

3

15	20	25	30				

Rule: ______

Working Mathematically

4 Draw the 3rd and 5th staircases in the pattern.
What's the next number of blocks in the pattern? ______

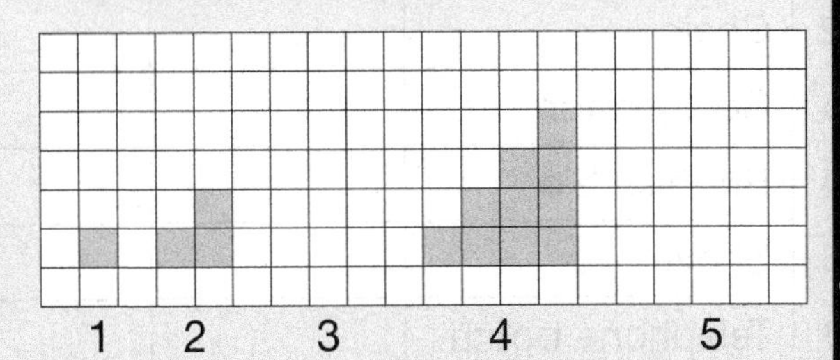

Number and Algebra

SET 3 Thirds

Shade the given fraction of each shape.

1 $\frac{1}{3}$

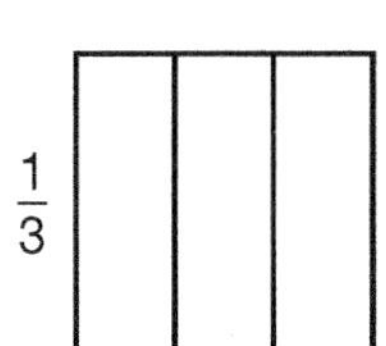

2 $\frac{2}{3}$

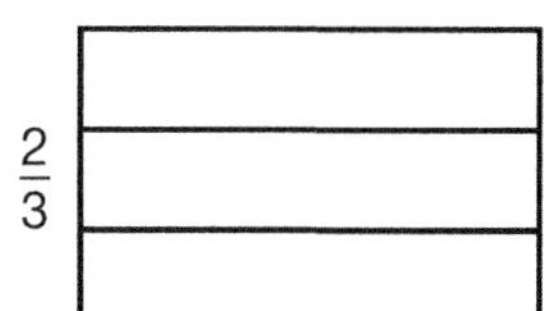

3 $\frac{1}{3}$

4 $\frac{2}{3}$

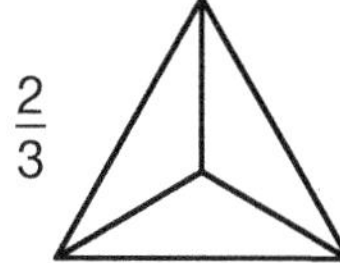

5 Shade $\frac{1}{3}$ of each group

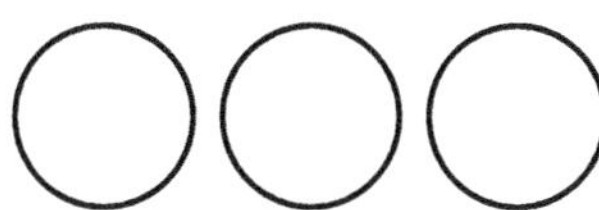

6

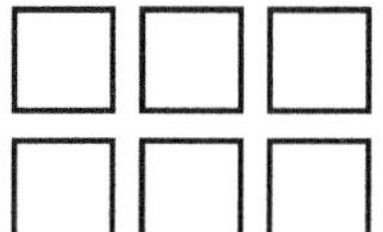

7 Is $\frac{1}{3}$ larger than $\frac{1}{2}$?

8 Is $\frac{1}{4}$ larger than $\frac{1}{3}$?

9 Is $\frac{1}{3}$ smaller than $\frac{1}{8}$?

SET 4 Extension

1 Half of 32

2 Put these numbers in counting order.

17, 12, 11, 19

3 23 + 35

4 Write 187 in words.

5 3 hundreds + 2 tens + 5 ones

6 How many cents in $1.75?

7 How many 10c coins in $1.50?

8 How much are 10 cakes at 15c each?

9 How many sides on 7 triangles?

10 How many centimetres in 2 m?

11 Twelve less than 58

12 Jane had $2 and spent $1.50. How much was left?

13 27 + ☐ = 40

14 36 = ☐ dozen

15 How many hours are between 11 am and 4 pm?

16 Name this shape.

Measurement Mass

Lift the following items, then order them from lightest to heaviest in each group by numbering them 1 to 4.

1

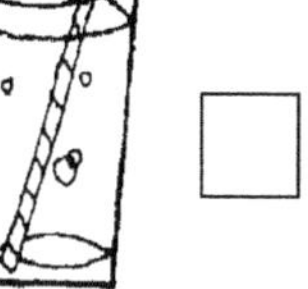

2

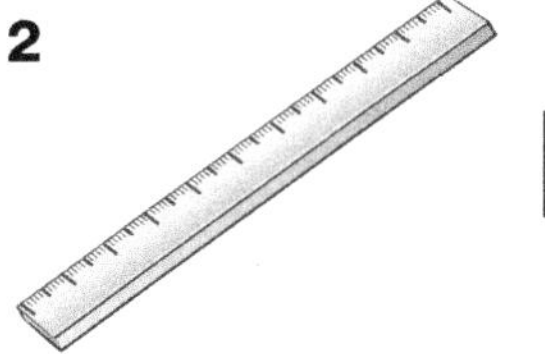

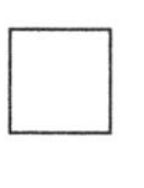

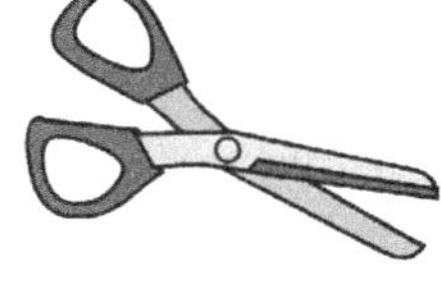

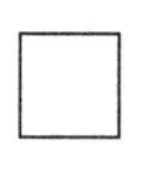

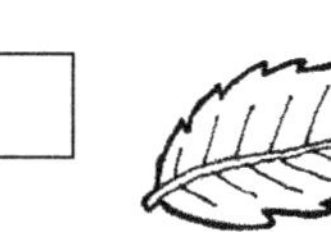

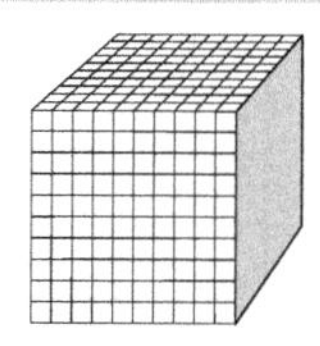

UNIT 12

Number and Algebra

SET 1 Basic

1 20 + 0

2 14 + 3

3 12 + 6

4 9 + 7

5 8 + 5

6 7 + 4

7 15 + 2

8 17 + 1

9 9 × 2

10 9 × 4

11 20 – 10

12 17 – 3

13 18 – 14

14 7 less than 17

15

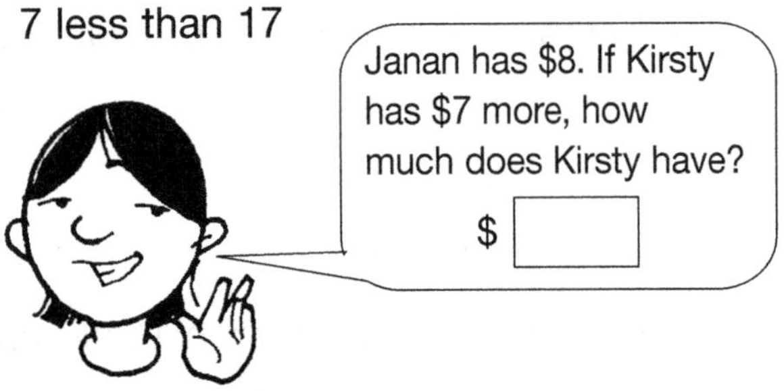
Janan has $8. If Kirsty has $7 more, how much does Kirsty have?
$

SET 2 Number lines

Put a cross on the number line to show the position of each number.

1 10

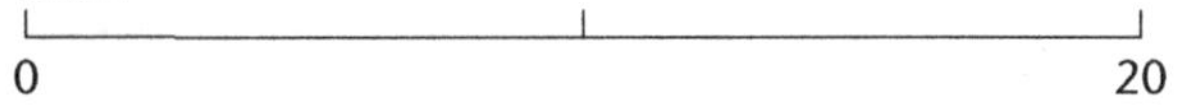
0
20

2 15

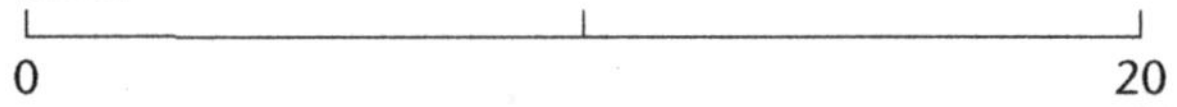
0
20

3 25

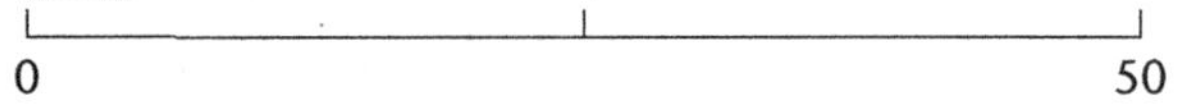
0
50

4 25

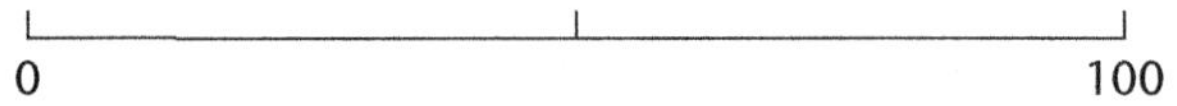
0
100

5 75

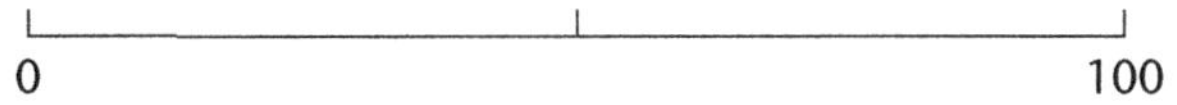
0
100

6 50

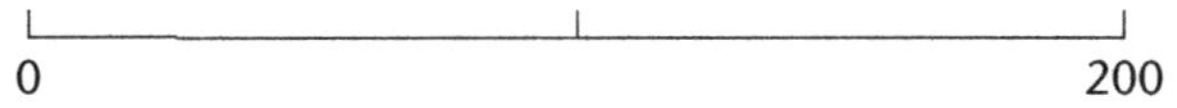
0
200

7 75

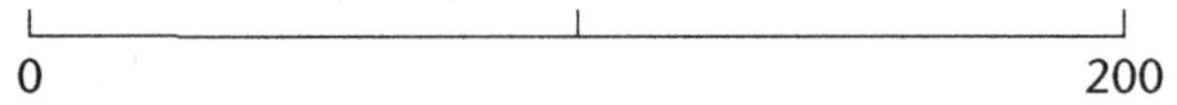
0
200

Space Symmetry

Use the line of symmetry to complete each shape.

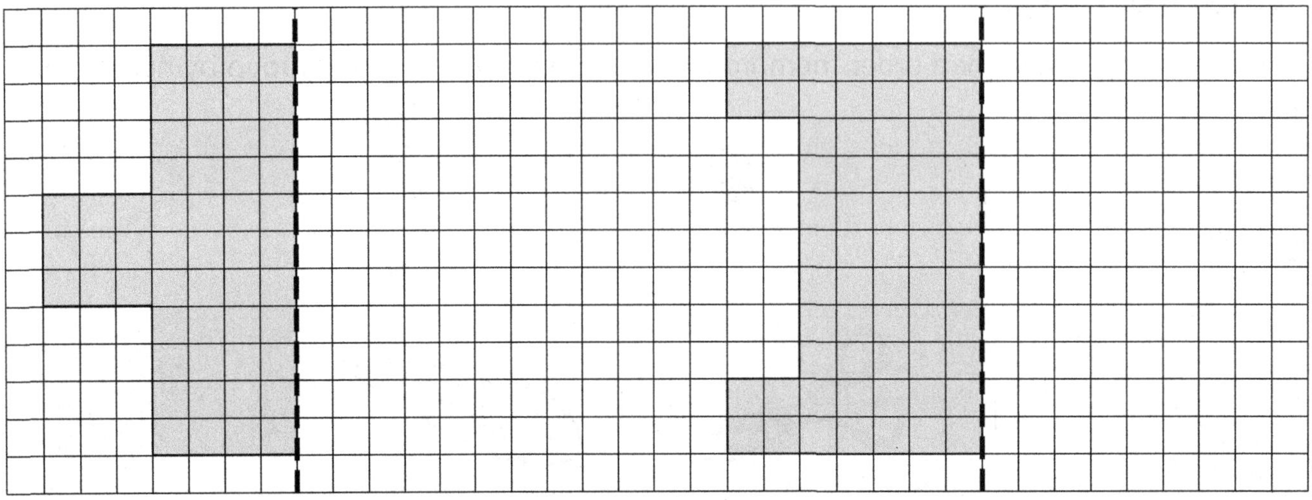

Number and Algebra

SET 3 Multiplication facts (10s)

1 10 lots of 3

2 10 rows of 5 plants

3 Multiply 4 by 10.

4 Product of 7 and 10

5 Ten lots of zero

6 How many wheels on 10 cars?

7 How many legs on 10 octopuses?

8

	3	5	6	2	9	8	7	4
×3	9			6				

9

	3	5	6	2	9	8	7	4
×2		10		4				8

10

	3	5	6	2	9	8	7	4
×10	30		60					

11

	3	5	6	2	9	8	7	4
×5	15				45		35	

SET 4 Extension

1 What is the hottest season?

2 Put these numbers in counting order.
7, 3, 2, 5, 1

3 $\frac{1}{2}$ of 26

4 How many tens in 96?

5 45 + 13

6 Write 64 in words.

7 8 tens and 9 ones

8 Share 36 among 3.

9 How many 20c coins in $3?

10 How much are 1 dozen chocolates at $2 each?

11 How many wheels on 7 cars?

12 How many centimetres in a metre?

13 How many cents in $1.20?

14 Fifteen less than 40

15 There are 35 children in Yr 3, 27 in Yr 4, 20 in Yr 5 and 32 in Yr 6. How many children are there altogether?

Statistics and Probability Chance

Sally had red shorts and blue shorts. She also had a red top and a blue top. Colour the clothes below to show the four combinations of colours that she could wear.

UNIT 13

Number and Algebra

SET 1 Basic

1 8 + 1 + 0

2 8 + 1 + 2

3 4×4

4 \$10 – \$6

5 12 + 7

6 13 + 5

7 20 + 0

8 7×3

9 18 – 5

10 25 – 20

11 19 minus 7

12 Ten less than twenty-four

13 Double 10.

14 Half of 12

15 

SET 2 Subtraction

	–	47	56	88	67	79
1	24	23				
2	35					

	–	79	98	65	99	87
3	44					
4	63					

5 97 children attended the camp. If 46 were boys, how many were girls? 

6 The school canteen made 76 lunches. 46 lunches were hamburgers and the rest were hot dogs. How many hot dogs were ordered?

Space Angles

1 Draw an angle smaller than a right angle.

2 Draw an angle larger than a right angle.

3 Colour the grid to classify the angles.

Angle	Smaller than a right angle	Right angle	Larger than a right angle

Number and Algebra

SET 3 Fifths and tenths

Colour the given fraction of each shape or group.

1 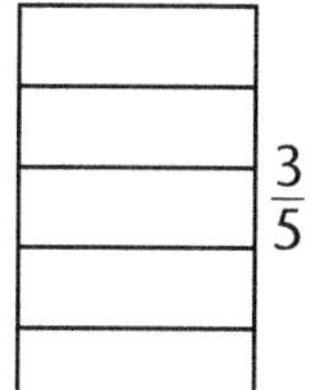 $\frac{3}{5}$

2 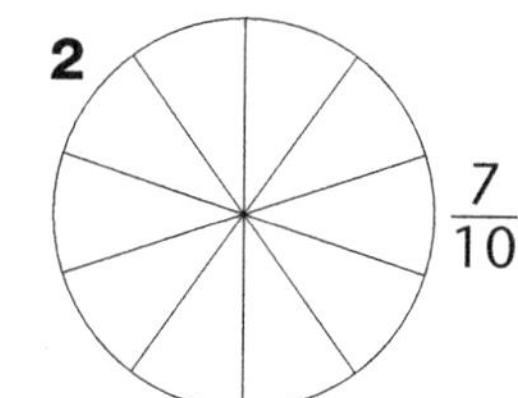 $\frac{7}{10}$

3 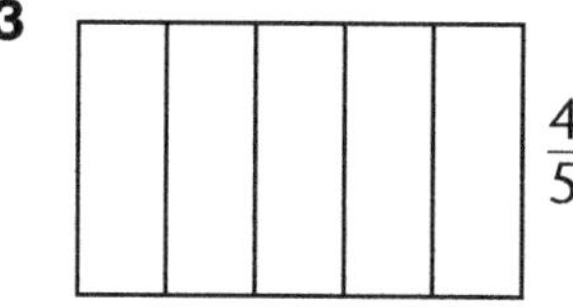 $\frac{4}{5}$

4 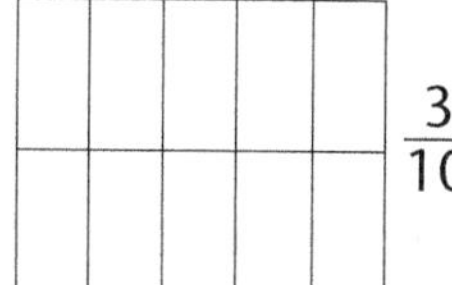 $\frac{3}{10}$

5  $\frac{2}{5}$

6 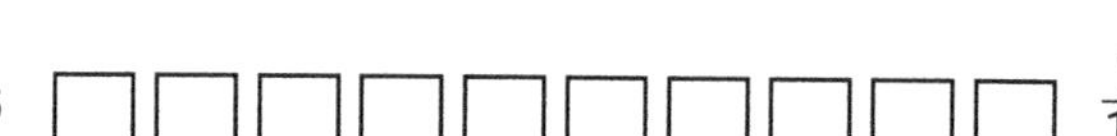 $\frac{9}{10}$

True or false?

7 $\frac{2}{5} < \frac{4}{5}$

8 $\frac{9}{10} > \frac{10}{10}$

9 $\frac{1}{2}$ of 20 = 10

10 $\frac{1}{5}$ of 10 = 2

SET 4 Extension

1 Fifty-two and thirty more

2 7 tens + 3 ones

3 How many sides has a triangle?

4 How many sides in 5 triangles?

5 \$28 – \$9

6 What is the number 20 more than 17?

7 52 – ☐ = 30

8 Three dozen minus twelve

9 Share 15 among 5.

10 How many 10c coins in \$2?

11 Which month comes before September?

12 Change from \$1 after spending 35c

13 Hours in $2\frac{1}{2}$ days

14 What is the product of 4 and 5?

15 If this month is June, what month will it be in 3 months' time?

16 Colour the third koala from the left.

Measurement Litres

Find the differences in capacity between the following:

1 the bucket and the ice-cream container. ☐

2 the cordial flask and the milk carton. ☐

3 the bucket and 2 full cordial flasks. ☐

4 the oil can and 4 lemonade bottles. ☐

5 How many times must the ice-cream container be filled in order to fill the bucket? ☐

6 How many full cordial flasks can be poured into the bucket? ☐

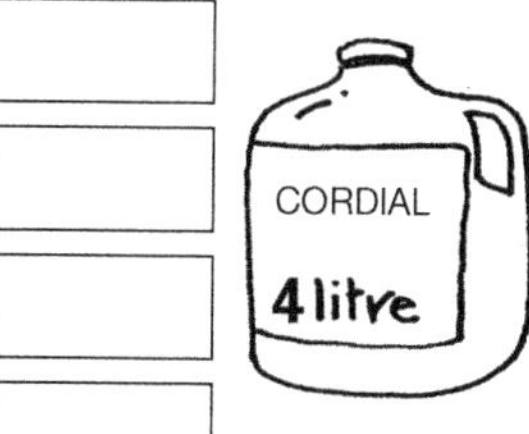

UNIT 14

Number and Algebra

SET 1 Basic

1 6 + 3 + 1

2 7 + 2 + 2

3 18 – 4

4 4 + 2 + 3

5 Two less than 7

6 5 + 2 + 5

7 Double 11.

8 8 + 5 + 4

9 10 + 2 + 3

10 One more than 18

11 9 + 3 + 2

12 Is 21 an odd number?

13 2 + 1 + 4

14 5 × 4

15 

Gina came 3rd in a race and Sam came 4 places behind Gina. Where did Sam come?

☐ th

SET 2 Bridging to decades

Add these numbers by first adding to 10.

	Numbers	Becomes	Answer
1	9 + 8	9 + 1 + 7	
2	18 + 7	18 + 2 + 5	
3	43 + 9		
4	64 + 8		
5	36 + 5		
6	45 + 7		

Use the road signs to work out the distances.

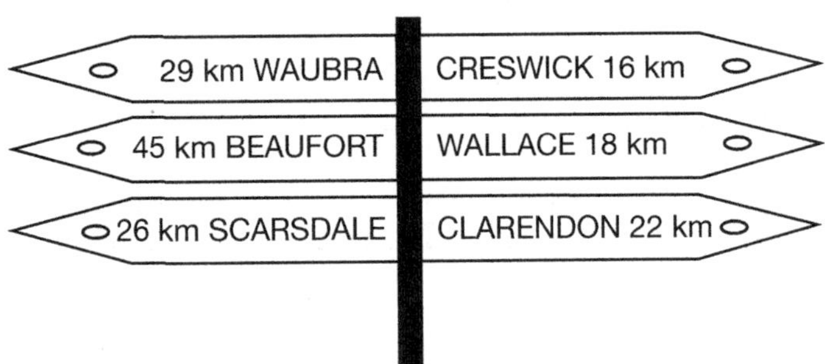

7	Wallace to Scarsdale	km
8	Waubra to Creswick	km
9	Beaufort to Clarendon	km
10	Creswick to Scarsdale	km
11	Wallace to Waubra	km

Space Three-dimensional objects

1 Colour the nets that would fold to make a cube.

a 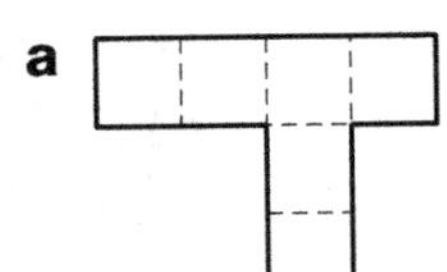

b 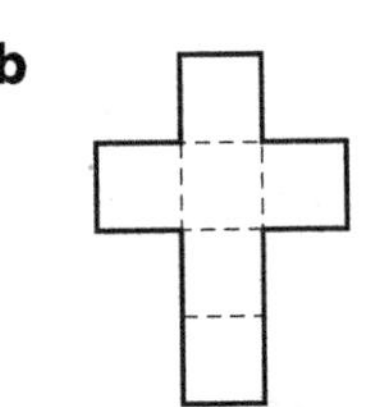

c 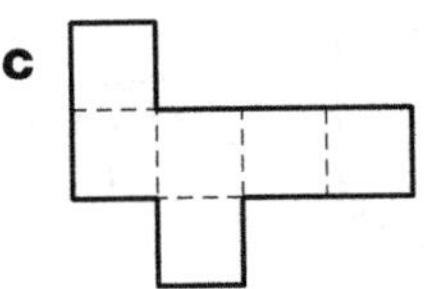

d 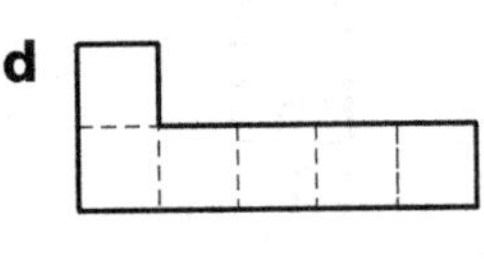

2 Count the number of cubes in each object.

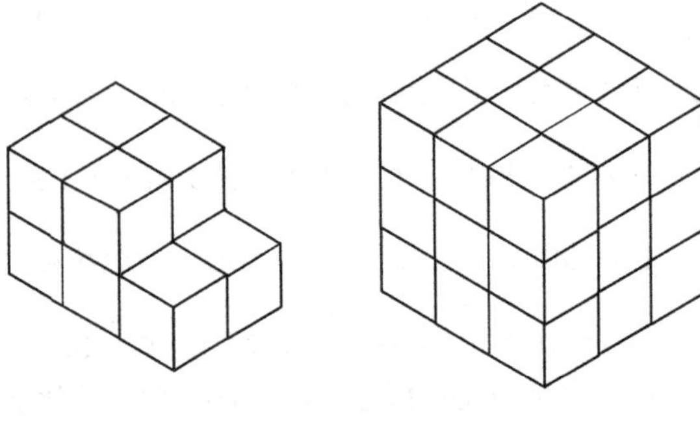

☐ cubes ☐ cubes

Number and Algebra

SET 3 Dividing

Use the stars to help solve the divisions.

☆ ☆ ☆ ☆ ☆ ☆
☆ ☆ ☆ ☆ ☆ ☆
☆ ☆ ☆ ☆ ☆ ☆
☆ ☆ ☆ ☆ ☆ ☆

1. 24 ÷ 4 = ☐
2. 24 ÷ 6 = ☐
3. 24 ÷ 2 = ☐
4. 24 ÷ 12 = ☐
5. 24 ÷ 8 = ☐
6. 24 ÷ 3 = ☐

Answer these questions.

7. 15 ÷ 3 = ☐
8. 15 ÷ 5 = ☐
9. 20 ÷ 4 = ☐
10. 25 ÷ 5 = ☐
11. 30 ÷ 6 = ☐

SET 4 Extension

1. Add 7 to 206.
2. How many tens in 300?
3. Write the number four hundred and six.
4. How many $2 coins make $50?
5. 37 – ☐ = 12
6. 56 + ☐ = 80
7. 276 = ☐ hund + ☐ tens + ☐ ones

Working Mathematically

Fill in the empty boxes to create number sentences that are true.

8. ☐ + ☐ = 16
9. ☐ + ☐ > 25
10. ☐ × ☐ > 23
11. ☐ ÷ ☐ = 4

Colour the two parts of each circle that multiply to give 28.

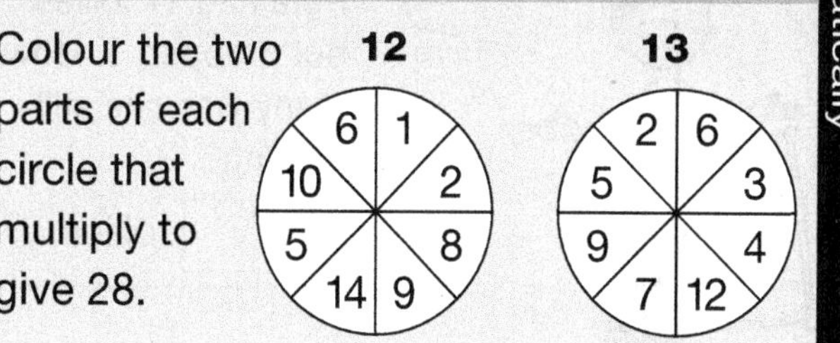

Statistics and Probability Two-way tables

Likes rugby	Likes soccer
Kim	Jessica
John	Kelly
Tim	Con
Greg	Abdul
Kelly	Tim
Maria	Sam
Soula	John

1. Which children like both rugby and soccer?

____

2. Which children like soccer only?

____

____

3. Which children like rugby only?

____

____

UNIT 15

Number and Algebra

SET 1 Basic

1 2 + 11

2 3 + 12

3 5 + 9

4 16 + 7

5 12 − 2

6 11 − 2

7 10 − 2

8 19 − 7

9 What is the sum of 12 and 8?

10 7 × 2

11 3 × 4

12 What is the product of 3 and 4?

13 Write the number that follows 324.

14 What is the sum of 3 and 10?

15

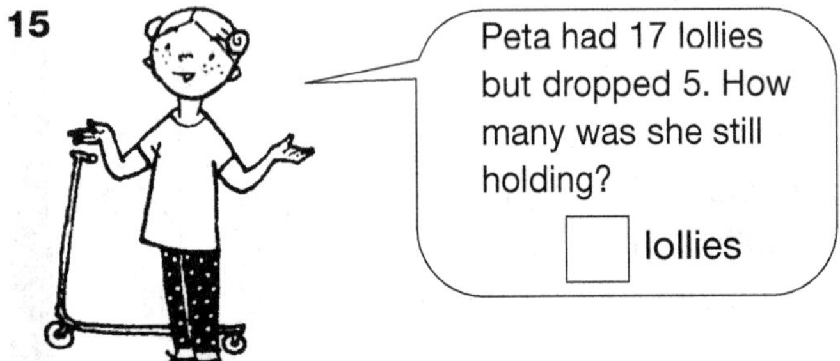

SET 2 Rounding numbers

Round these numbers to the nearest 10 before adding them to make an estimate.

1 39 + 21 = ____

2 58 + 19 = ____

3 43 + 49 = ____

4 38 + 42 = ____

5 51 + 37 = ____

6 69 + 22 = ____

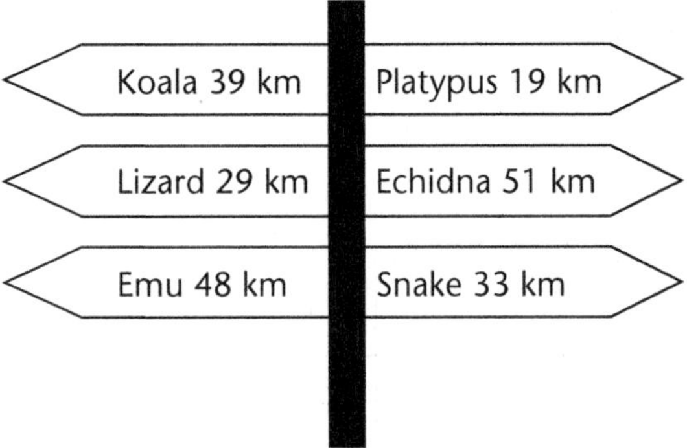

Estimate the distances from:

7	Koala to Platypus	km
8	Lizard to Echidna	km
9	Emu to Snake	km

Number and Algebra Doubling and halving

Complete the function machines.

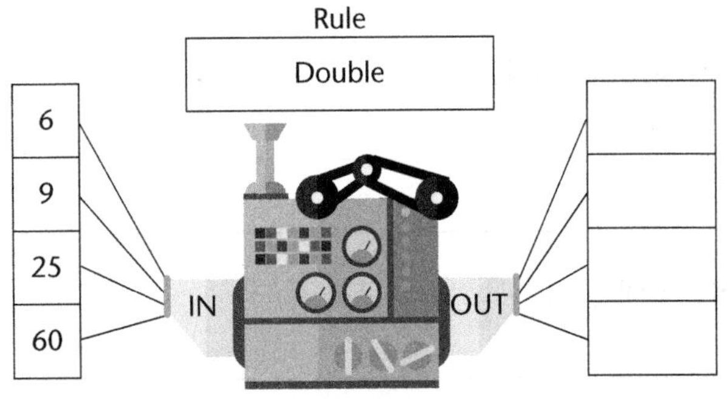

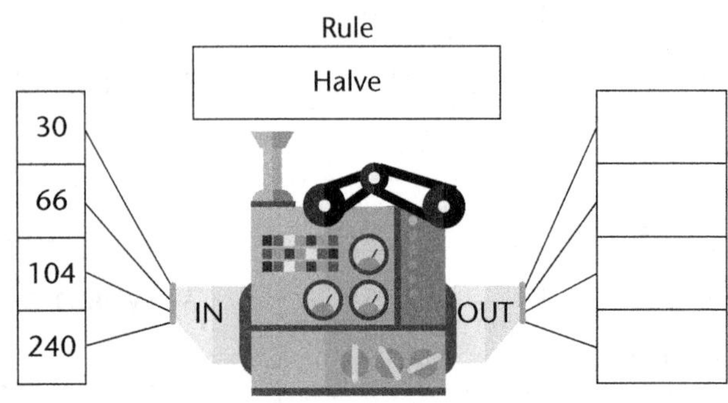

Number and Algebra

SET 3 4-digit numbers

Show each number on the number expander.

1 2376

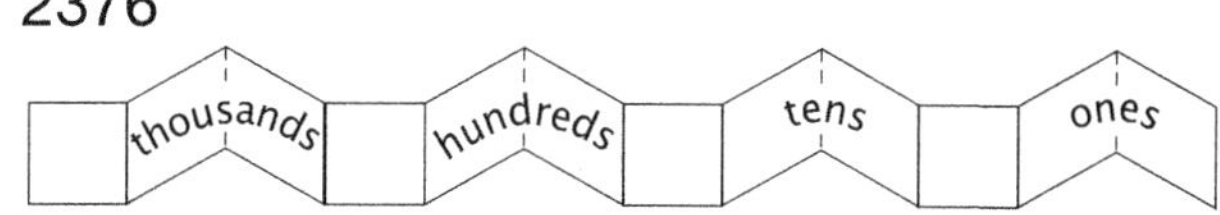

2 3591

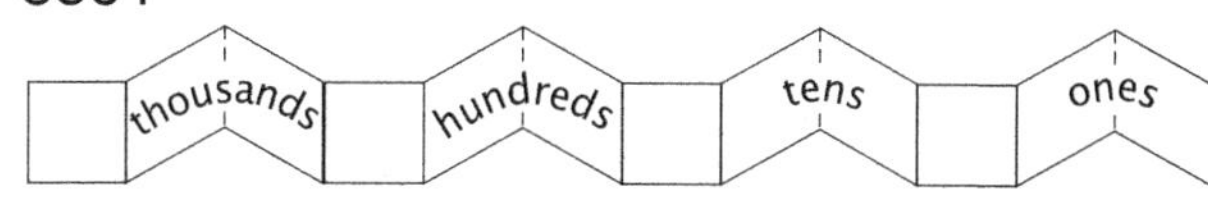

3 4385

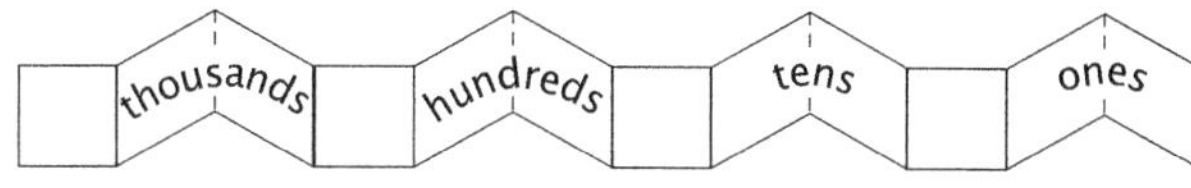

4 3296

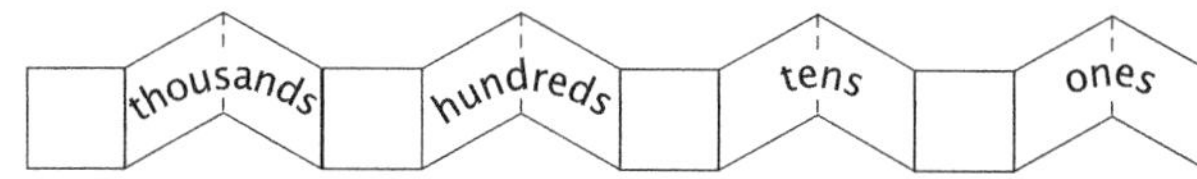

Write these numbers in expanded form.

5 2579 = 2000 + 500 + ☐ + ☐

6 3628 = ☐ + ☐ + ☐ + ☐

7 4196 = ☐ + ☐ + ☐ + ☐

8 7896 = ☐ + ☐ + ☐ + ☐

9 9494 = ☐ + ☐ + ☐ + ☐

10 Order these numbers from smallest to largest.

1437, 3471, 1341 ☐

SET 4 Extension

1 Add 14 to 31.

2 What is the product of 4 and 6?

3 How many days in 3 weeks?

4 Estimate an answer to 37 + 69.

5 Round 363 to the nearest 10.

6 65 + ☐ = 80

7 Total of 13 and 12

8 3 hundreds + 2 tens + 7 ones

9 Share 40 among 5.

10 How many centimetres in 3 m?

11 Write 172 in words.

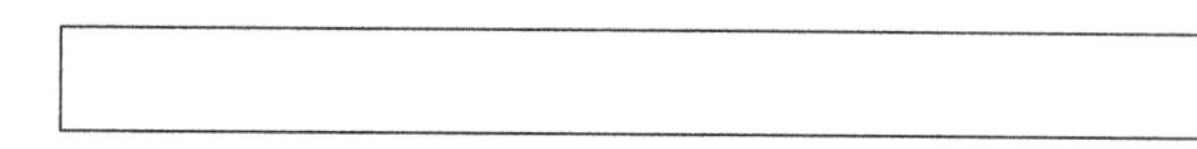

12 Take 45 from 85.

13 48 hours = ☐ days

14 70c + 5c + 25c

Working Mathematically

How old are Rosie and Dev if:

15 Rosie is a teenager whose age is a multiple of 7?	
16 Dev is a teenager whose age is a multiple of 5?	

Measurement Time

Add the minute hands to these clocks.

1 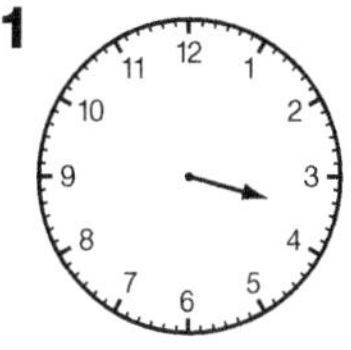

Half past 3

2 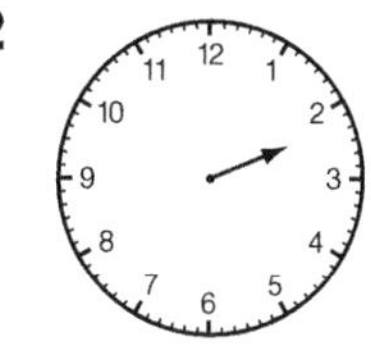

A quarter past 2

3 

Half past 8

4 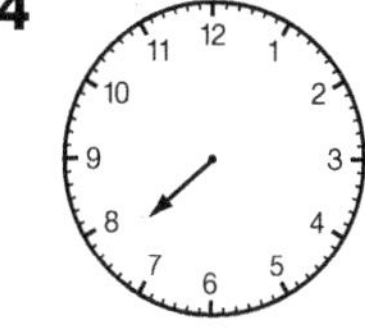

A quarter to 8

5 

Half past 10

UNIT 16

Number and Algebra

SET 1 Basic

1 17 + 6

2 19 − 3

3 5 × 10

4 16 + 4

5 26 − 4

6 4 × 5

7 17 + 3

8 17 − 3

9 5 × 4

10 Product of 5 and 3

11 Sum of 5 and 10

12 Product of 4 and 5

13 Difference of 8 and 4

14 7 less than 9

15

Luke had $28 but spent $8 on a toy. How much money has he left?

$ ☐

SET 2 Addition with magic squares

The sum of each vertical, horizontal and diagonal line of numbers is the same.

Example

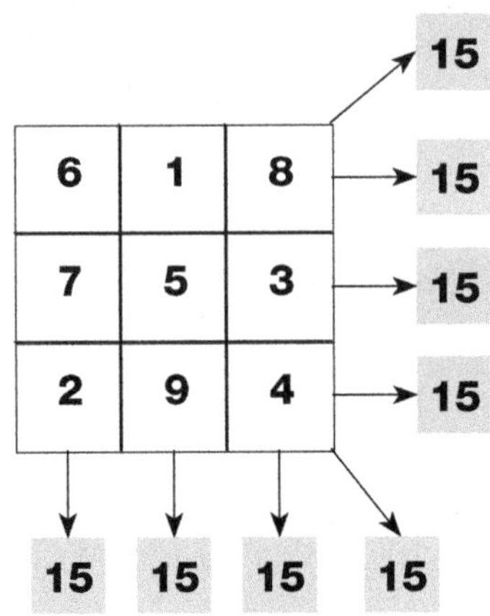

Complete these magic squares.

1

9	4	5
		10
7		3

2

7	2	9
	6	4
	10	

3

	4	11
10	8	6
	12	

4

14	9	10
		15
12		8

Statistics and Probability Collecting data

Our class did a survey on the type of pets we had at home. Colour the boxes in the graph to record the pets tallied in our survey.

Pet survey

Dog

Cat

Bird

Fish

Mouse

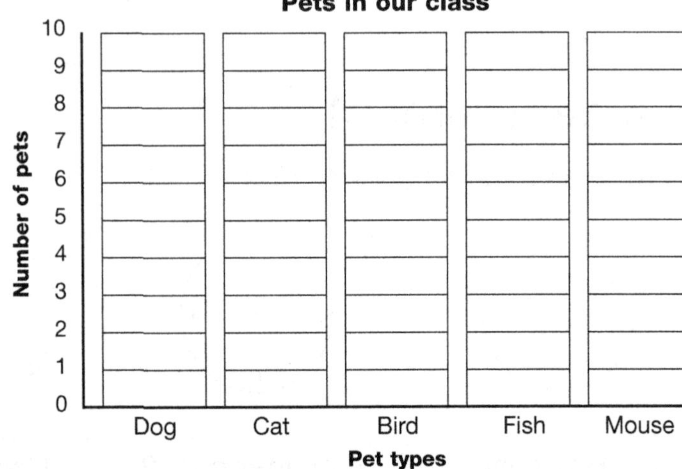

Number and Algebra

SET 3 Multiplication facts (2s, 5s, 10s)

1 3 × 2
2 5 × 2
3 7 × 2
4 9 × 2
5 6 × 2
6 3 × 10
7 5 × 10
8 7 × 10
9 9 × 10
10 6 × 10
11 3 × 5
12 5 × 5
13 7 × 5
14 9 × 5
15 6 × 5
16 8 × 5

17 How many different arrays can you make with 24 counters? ______
Draw one of your arrays.

Working Mathematically

SET 4 Extension

1 Round 1328 to the nearest 100.
2 How many sides in 14 rectangles?
3 95 = ☐ tens + ☐ ones
4 162 = ☐ hund + ☐ tens + ☐ ones
5 Total of 27 and 600
6 Share 60 among 6.
7 13 more than 122
8 Write 39 in words. ☐
9 What do I add to 35 to make 50?
10 150, ☐ , 250, ☐ , 350
11 How much are 5 envelopes at 50c each?
12 What odd number comes before 240?
13 Write 7 as an ordinal number.
14 How many hundreds in 375?
15 Jing bought 8 lollies for 10c each. How much change did she get from $2? 

Measurement The kilogram

Colour the correct sets of scales.

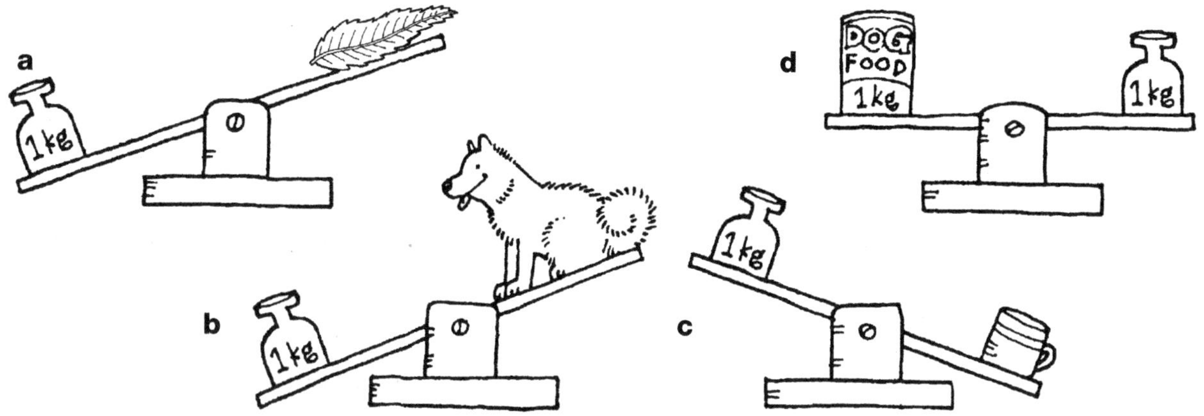

UNIT 17

Number and Algebra

SET 1 Basic

1 6 + ☐ = 10

2 4 times 4

3 Double 6.

4 14 – 10

5 6 × 10

6 5 × 10

7 17 – ☐ = 14

8 17 – ☐ = 7

9 15 + 4

10 What day follows Thursday?

11 10 + 16

12 10, 20, ☐ , ☐ , 50

13 Complete the grid to show how much Jenny will save in 5 weeks.

Weeks	1	2	3	4	5
Amount	$3	$6	$9		

SET 2 Empty number lines

Model and answer the additions and subtractions on the empty number lines. The first one has been done for you.

1 37 – 20 = 17

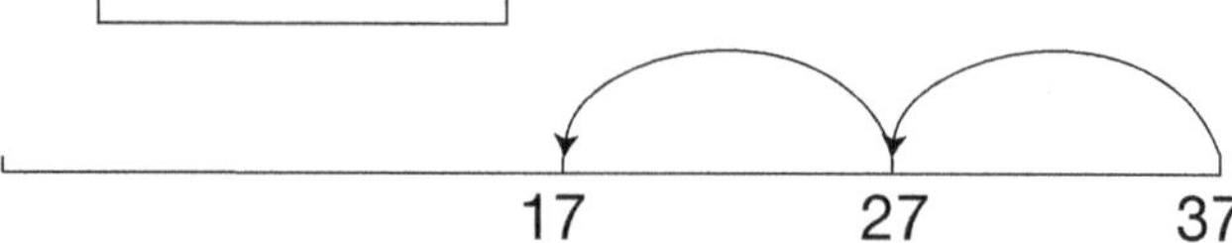

2 68 + 14 =

68

3 77 – 13 =

77

4 38 + 14 =

38

5 75 – 14 =

75

Space Following directions

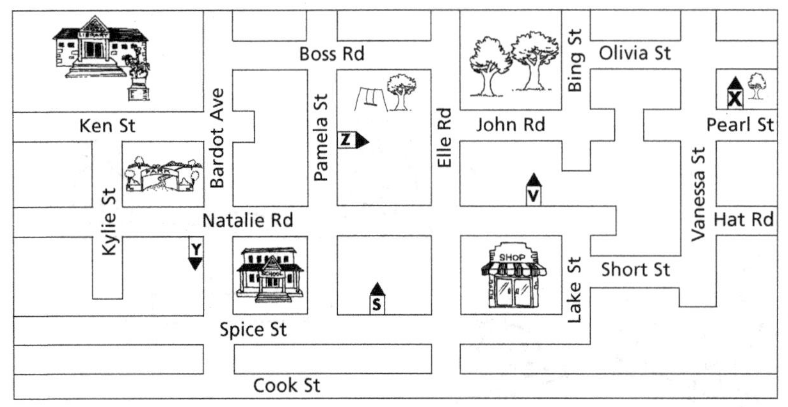

Draw the path on the map.

1 Start at the X and travel along Pearl St until it meets Vanessa St.

2 Turn left into Vanessa St, then right into Short St.

3 Turn left into Lake St then right into Spice St.

4 Follow Spice St until it meets Bardot Ave, then turn right.

5 Follow Bardot Ave until it meets Natalie Rd.

6 What letter did you find?

Number and Algebra

SET 3 Division and multiplication

1 Four tickets were bought and each ticket cost $7. How much were the tickets in total? 

2 Three tickets were bought and each ticket cost $8. How much were the tickets in total? 

3 Complete the grid.

÷	24	32	16	12	8	40	4
4	6			3			

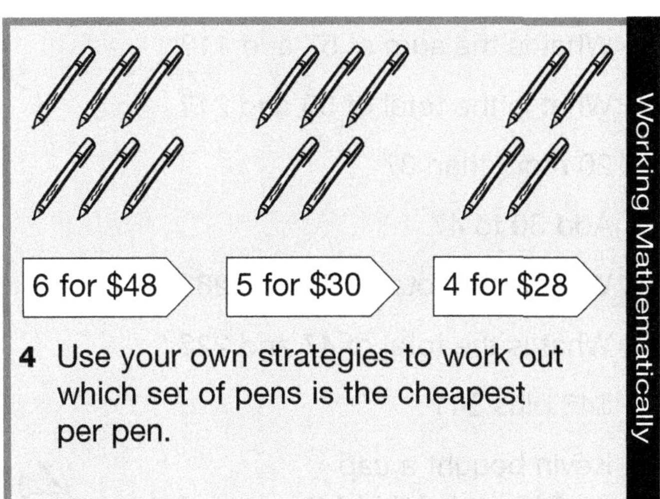

4 Use your own strategies to work out which set of pens is the cheapest per pen.

SET 4 Extension

1 What is the sum of 3, 6 and 8?

2 1 litre = ☐ millilitres

3 Which is larger, 2 dozen or 21?

4 Write the largest number you can, using 4, 7, 5.

5 Round 2326 to the nearest 100.

6 How many minutes are in half an hour?

7 4 times zero

8 Estimate an answer to 136 + 141.

9 How many tens are in 894?

10 What is the product of 6 and 2?

11 Share 30 among 6.

12 137, 143, 149, ☐ , ☐

13 How much are 2 tickets at 75c each?

14 Twenty centimetres plus 130 cm

15 Circle the date 21 days after 17 May.

MAY

Su	M	Tu	W	Th	F	S
	1	2	3	4	5	6
7	8	9	10	11	12	13
14	15	16	17	18	19	20
21	22	23	24	25	26	27
28	29	30	31			

JUNE

Su	M	Tu	W	Th	F	S
				1	2	3
4	5	6	7	8	9	10
11	12	13	14	15	16	17
18	19	20	21	22	23	24
25	26	27	28	29	30	

Measurement Perimeter

1 Measure the perimeter of this shape.

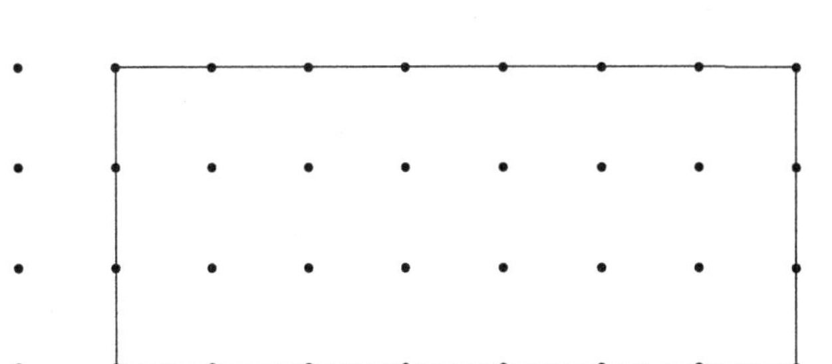

The perimeter is ______ cm.

2 Draw a shape with a perimeter of 12 cm.

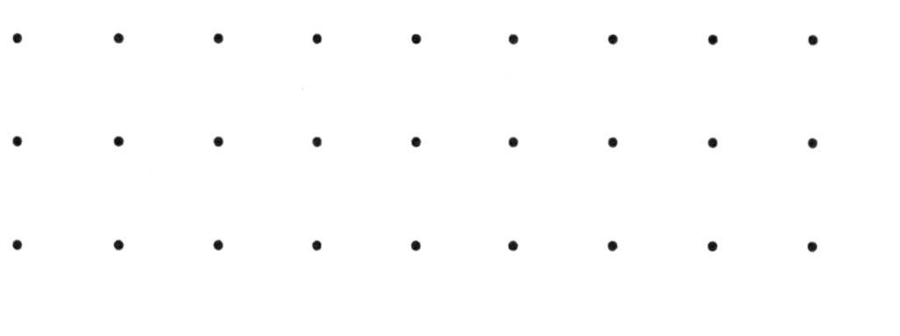

Number and Algebra

SET 1 Basic

1 5 + 3

2 15 + 3

3 25 + 3

4 35 + 3

5 12 – 5

6 16 – 5

7 16 – 6

8 16 – 8

9 16 – 10

10 7 × 4

11 9 × 5

12 8 × 4

13 3 × 4

14 Product of 7 and 5

15

John saved $4 each week for 7 weeks. How much did he save?

$ ☐

SET 2 Trading in addition

1

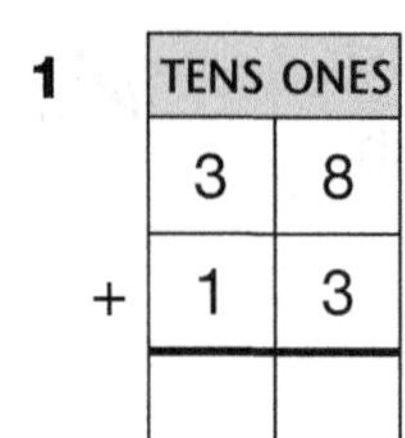

	TENS	ONES
	3	8
+	1	3

2

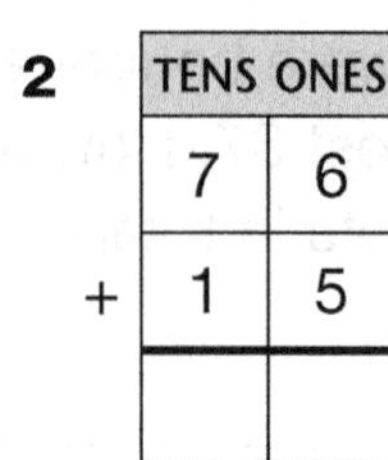

	TENS	ONES
	7	6
+	1	5

3

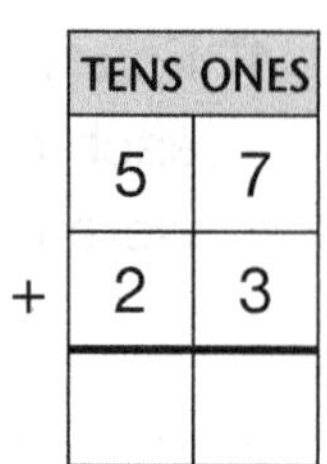

	TENS	ONES
	5	7
+	2	3

4

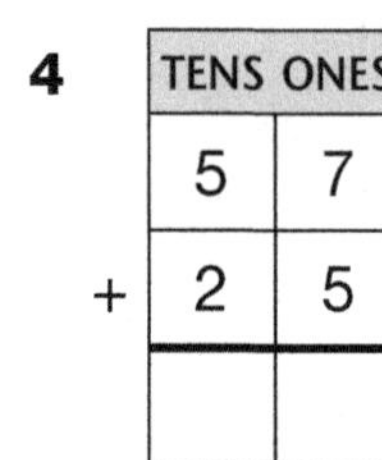

	TENS	ONES
	5	7
+	2	5

5

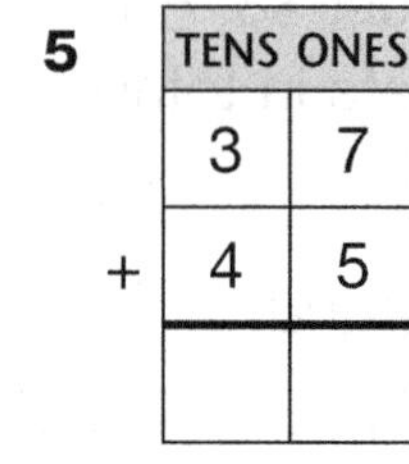

	TENS	ONES
	3	7
+	4	5

6

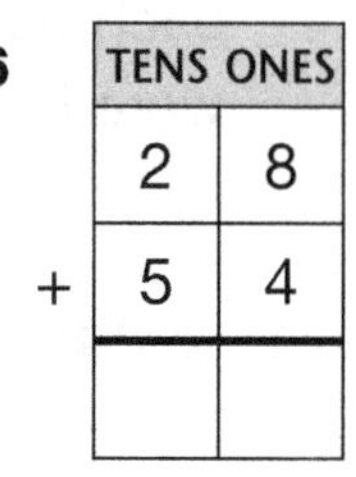

	TENS	ONES
	2	8
+	5	4

7 Add 63 and 10.

8 24 more than 54

9 What is the sum of 57 and 11?

10 What is the total of 56 and 21?

11 20 more than 37

12 Add 30 to 47.

13 What is the total of 53 and 26?

14 What is the total of 47 and 22?

15 $48 plus $41

16 Kevin bought a cap for $15 and a shirt for $38. How much did he spend altogether?

Space Describing prisms

Draw a line to match the prisms to their sets of faces.

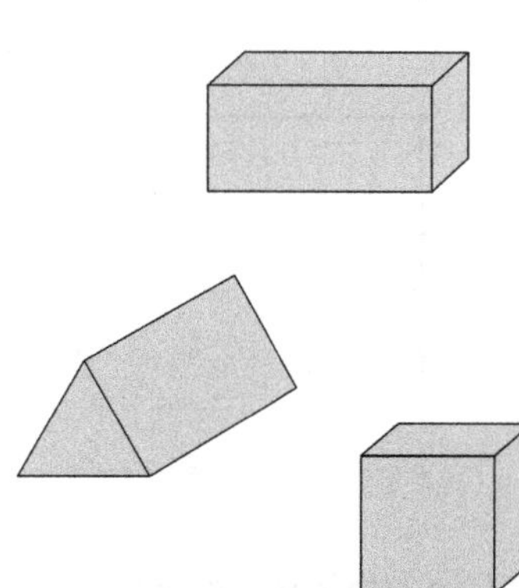

Number and Algebra

SET 3 Number patterns

Discover the rules and complete the grids.

1	$75	$70	$65			
2	$16	$20	$24			

Complete the patterns.

3	0	7	14	21				
4	62	58	54	50				
5	0	9	18	27				
6	57	65	73	81				

Complete each pattern then write the rule for it.

7	20	30	40	50				

Rule:

8	9	12	15	18				

Rule:

SET 4 Extension

1 How many hundreds are there in 866?

2 $\frac{1}{4}$ of 12

3 How many hours from 9 am to noon?

4 What is the difference between 47 and 23?

5 Paco bought 5 books at $7 each. What was the total cost?

6 4 hundreds + 7 tens + 3 ones

7 Write the largest 3-digit number you can using 7, 9, 1.

8 Write the smallest 3-digit number you can using 7, 9, 1.

9 936 + 4 tens

10 If juice costs $2 a litre, how much would half a litre cost?

Working Mathematically

11 Ms Stevens gives the students in our class 1 sticker each during the week. She gave out 16 stickers on Monday, but only half that number on Tuesday. She continued this pattern for the rest of the week. How many students are in our class?

Measurement The square centimetre

Draw 2 more shapes that cover an area of 12 cm^2.

UNIT 19

Number and Algebra

SET 1 Basic

1 9 + 4

2 $10 – $6

3 Half of 16

4 $7 + $9

5 Subtract 6 from 11.

6 4 × 5

7 15 – 3

8 7 less than 18

9 5 × ☐ = 15

10 6 × ☐ = 24

11 Double 12.

12 Sum of 13 and 6

13 21 + 4 + 10

14 Product of 10 and 9

15

Joe had 14 marbles but lost half of them. How many has he left?
☐ marbles

SET 2 Counting by hundreds

Complete the counting patterns.

1 | 300 | 400 | 500 | | | |

2 | 150 | 250 | 350 | | | |

3 | 900 | 800 | 700 | | | |

4 | 233 | 333 | 433 | | | |

5 | 880 | 780 | 680 | | | |

6 | 171 | 271 | 371 | | | |

Complete the grids.

7

+	79	230	110	803	600
100					

8

–	300	512	120	376	988
100					

9

+	5	105	205	305	405
100					

Space Reflect and rotate

1 Continue the reflection pattern.

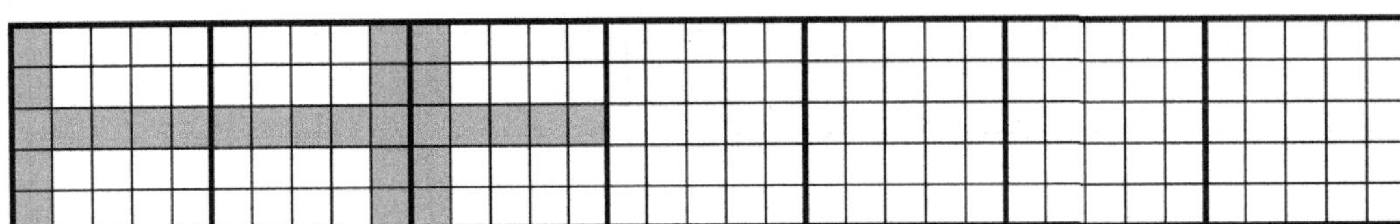

2 Continue the rotation pattern.

Number and Algebra

SET 3 Multiplication facts (3s)

1

× 5	
4	
6	
5	
7	
9	
8	

2

× 3	
3	
7	
6	
8	
9	
5	

3

× 2	
1	
2	
4	
9	
6	
5	

Complete the algorithms to identify the winning bingo card.

4 $3 \times 3 =$

5 $6 \times 3 =$

6 $4 \times 3 =$

7 $5 \times 3 =$

8 $7 \times 3 =$

9 $8 \times 3 =$

10 $6 \times 5 =$

11 $9 \times 5 =$

a

21		15		48	
	30		18		24

b

21		9		15	
	12		18		45

c

21		45		12	
	18		17		48

SET 4 Extension

1 Cost of a $10 cake and a $6 tart

2 How many days in 5 weeks?

3 What is the next odd number after 49?

4 How many minutes from 10 am to 2 pm?

5 Share 60 among 5.

6 $72 + $23

7 6 hundreds plus 7 tens plus 5 ones

8 34 tens plus 9 ones

9 How many 5c coins make 35c?

10 223, 229, 235, ☐, ☐

Working Mathematically

11 Jenna has forgotten her 3-digit PIN. The numbers are 4, 5 and 6.

Write all 6 possible combinations on the cards.

My Bank	My Bank	My Bank
☐☐☐	☐☐☐	☐☐☐
My Bank	**My Bank**	**My Bank**
☐☐☐	☐☐☐	☐☐☐

Statistics and Probability Chance

Shade the tag which best describes the chance of the spinner landing on blue.

1

red
blue
green
white

○ NEVER

○ UNLIKELY

○ CERTAIN

○ LIKELY

Design a spinner with the colours red, blue and yellow on it. Make sure that red has a higher chance of being landed on than blue.

2

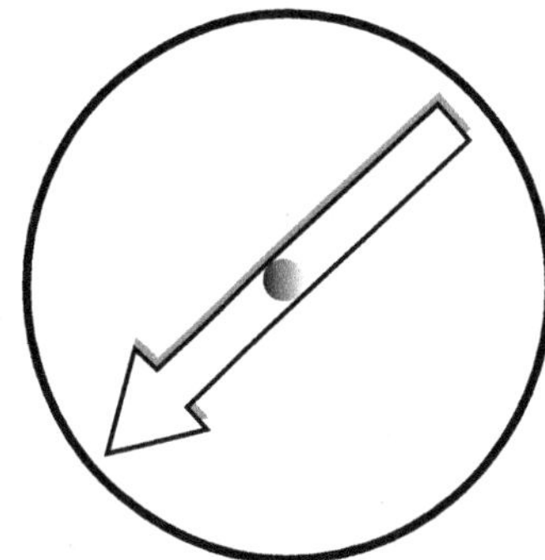

Number and Algebra

SET 1 Basic

1 23 + 4

2 24 + 5

3 25 + 7

4 27 + 8

5 58 minus 4

6 19 – 5

7 14 – 4

8 18 – 4

9 5 × 3

10 4 × 3

11 3 × 9

12 2 × 6

13 10 × 9

14 7 tens and 5 ones

15 Lauren had $27 but gave $13 to Kelly. How much does she have left?

$ ☐

SET 2 Trading in subtraction

1

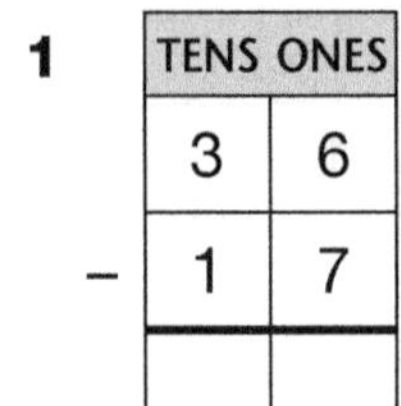

	TENS	ONES
	3	6
–	1	7

2

	TENS	ONES
	4	8
–	1	9

3

	TENS	ONES
	5	2
–	3	7

4

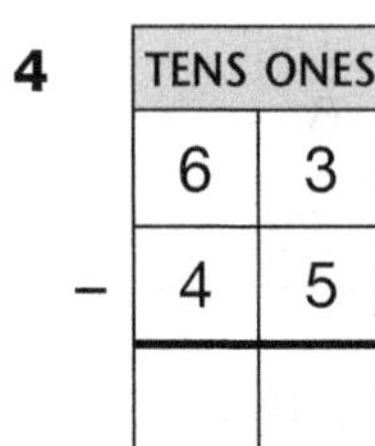

	TENS	ONES
	6	3
–	4	5

5

	TENS	ONES
	5	5
–	3	8

6

	TENS	ONES
	7	7
–	2	9

7

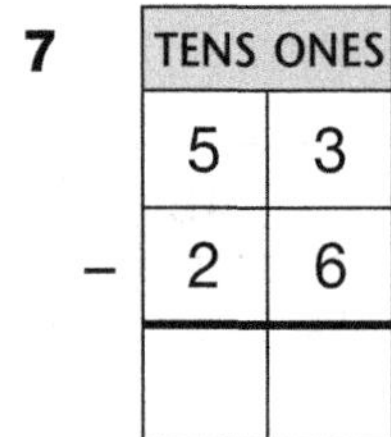

	TENS	ONES
	5	3
–	2	6

8

	TENS	ONES
	6	5
–	4	8

9

	TENS	ONES
	7	4
–	1	6

10

	TENS	ONES
	6	2
–	2	5

11

	TENS	ONES
	5	1
–	2	6

12

	TENS	ONES
	6	3
–	2	9

Space Angles

Draw a line to match each angle on the left to one on the right.

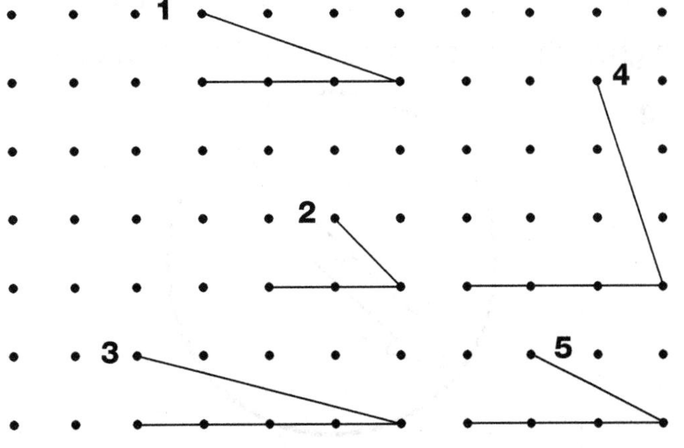

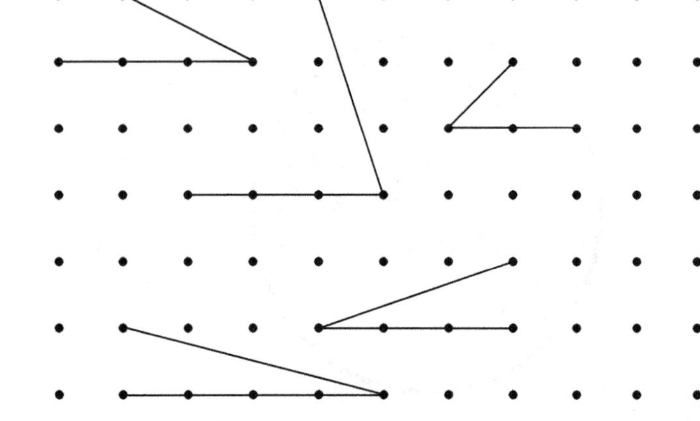

Number and Algebra

SET 3 Division from multiplication

Solve the questions below to find the secret words.

1 3 × 8 = ☐

2 24 ÷ 3 = ☐

3 5 × 8 = ☐

4 40 ÷ 5 = ☐

5 ☐ × 3 = 30

6 4 × 10 = ☐

7 7 × 3 = ☐

8 21 ÷ 3 = ☐

24	7	2	40	9
I	M	T	E	P

10	8	3	21	12
R	C	S	A	L

9

___ ___ ___   ___ ___ ___ ___ ___

1 2 3   4 5 6 7 8

10 Joe had 27 marbles which he needed to break into groups of 9. How many groups could he make?

11 Jill the tiler needed to put 40 tiles into groups of 9. How many groups did she have and how many left over?

SET 4 Extension

1 What is the value of 6 in 363?

2 How many hundreds in 573?

3 How many days in 8 school weeks?

4 473c = $ ☐

5 How much more than 130 is 210?

6 What are the factors of 12?

7 Write two thousand, seven hundred and twenty in figures.

8 What is the cost of 3 kg of potatoes at $3 per kg?

9 Jan bought 10 lollies at 17c each. She spent $ ☐

10 30 + 30 + ☐ = 170

11 Write 4 number sentences that equal 75.

☐ + ☐

☐ + ☐

☐ − ☐ = 75

☐ − ☐

Working Mathematically

Measurement Millimetres

Change the measurements from centimetres into millimetres.

1 6 cm = ______ mm

2 9 cm = ______ mm

3 4 cm = ______ mm

4 5 cm = ______ mm

5 2 cm = ______ mm

6 8 cm = ______ mm

7 10 cm = ______ mm

8 7 cm = ______ mm

9 1 cm = ______ mm

10 $\frac{1}{2}$ cm = ______ mm

UNIT 21

Number and Algebra

SET 1 Basic

1 7 + ☐ = 10

2 Add 7 to 11.

3 Subtract 10 from 50.

4 24 – 4

5 The difference between 47 and 5

6 5 lots of 3

7 7 × 6

8 8 × 4

9 25c + 5c + ☐ = 40c

10 How many hundreds in 224?

11 53 + 6

12 Product of 8 and 5

13 Sum of 8 and 5

14 Days in August

15

SET 2 3-digit addition

1 Add 27 to 108.

2 What is the sum of 32 and 27?

3 What is 57 more than 85?

4 The sum of 54 and 72

5 What is the cost of a $95 pair of shoes and a $44 suit?

6 What is the total of 25, 37 and 99?

7 Add 43 to 73.

8

	HUND	TENS	ONES
	1	7	3
+	2	2	5

9

	HUND	TENS	ONES
	4	3	6
+	5	5	3

10

	HUND	TENS	ONES
	4	0	7
+	4	6	8

11

	HUND	TENS	ONES
	2	1	6
+	3	0	7

12

	HUND	TENS	ONES
	4	0	5
+	3	0	7

13

	HUND	TENS	ONES
	3	5	7
+	2	1	6

Number and Algebra Evaluating strategies

Use any strategy you wish to answer the additions.

1 34 + 46 = ☐

2 49 + 51 = ☐

3 27 + 33 = ☐

4 36 + 26 = ☐

5 47 + 35 = ☐

6 85 + 97 = ☐

7 122 + 135 = ☐

8 182 + 211 = ☐

9 135 + 288 = ☐

Number and Algebra

SET 3 Calculating change

Calculate the change from $10 for each item

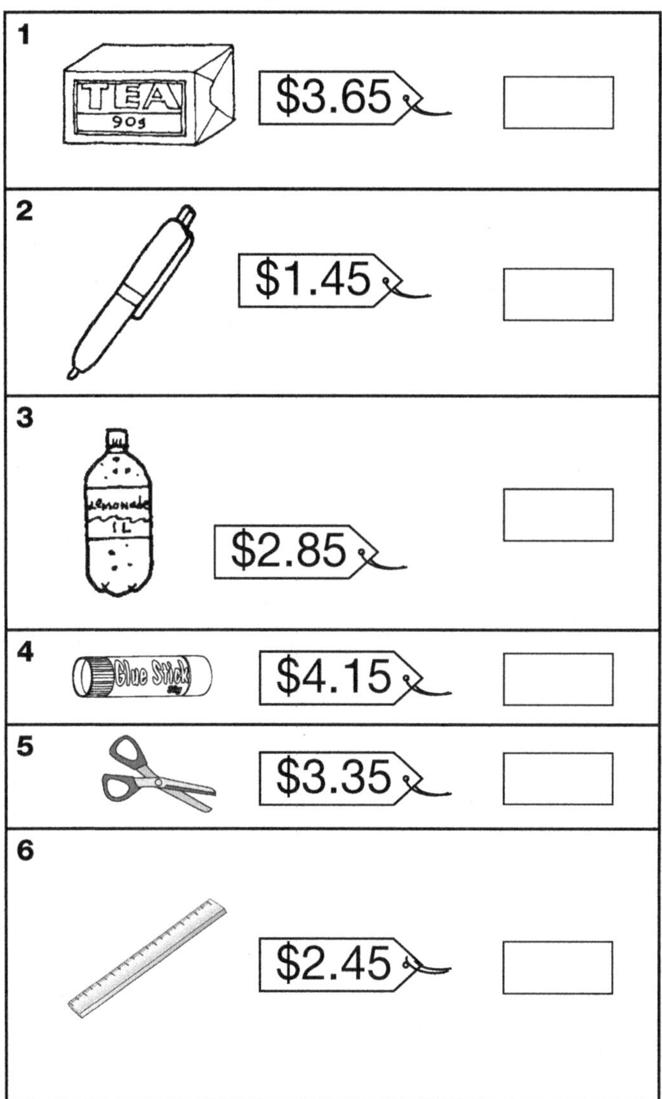

SET 4 Extension

1. ☐ × 15 = 30
2. Taj spent 5c and 15c. If he had 20c, how much has he left?
3. How many halves in 3 wholes?
4. 78, 81, 84, ☐, ☐
5. 3 cakes cost $12. How much did each cake cost?
6. 3 boys shared $60. How much did each boy get?
7. Is a cube a 3D object?
8. How many 10c coins in $2.30?
9. What is the value of 7 in 7323?
10. Round 2354 to the nearest 1000.
11. How many centimetres in 3 m?
12. If eggs cost $1.20 a dozen, how much for $\frac{1}{2}$ dozen?
13. What is $\frac{1}{4}$ of 48 lollies?
14. Write the even numbers between 231 and 239.

Measurement Litres

List items that have their capacity measured in litres.

Item	Capacity
Milk	1 litre

UNIT 22

Number and Algebra

SET 1 Basic

1 11 – 2

2 28 – 4

3 36 – 5

4 47 – 6

5 9 + 8 + 7

6 3 + 4 + 3

7 26 take away 7

8 7 + 3 + 8

9 4 × 6

10 3 × 6

11 5 × 6

12 10 × 6

13 18 plus 6

14 Sum of 17 and 12

15 

Petra had 6 red lollies, 9 green ones and 6 blue ones. How many did she have altogether?

[ ] lollies

SET 2 Trading in subtraction

1

	HUND	TENS	ONES
	8	4	3
–		2	4

2

	HUND	TENS	ONES
	2	9	2
–		3	4

3 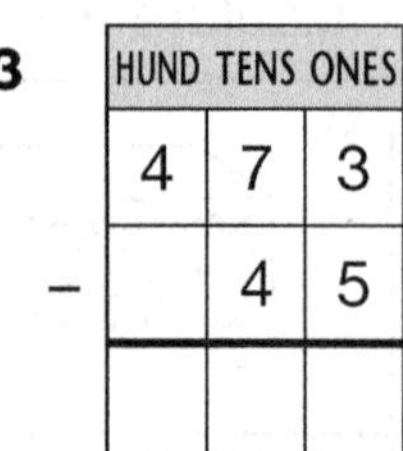

	HUND	TENS	ONES
	4	7	3
–		4	5

4

	HUND	TENS	ONES
	6	8	2
–		4	6

5

	HUND	TENS	ONES
	7	7	1
–		5	4

6 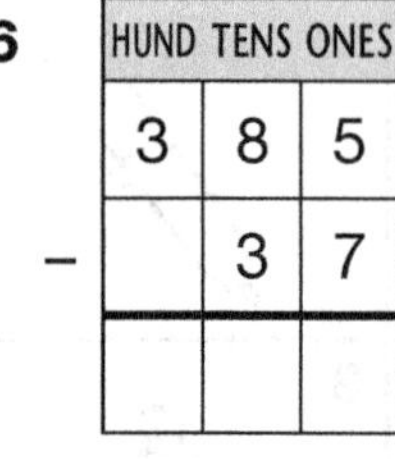

	HUND	TENS	ONES
	3	8	5
–		3	7

7

	HUND	TENS	ONES
	4	9	4
–	2	3	7

8 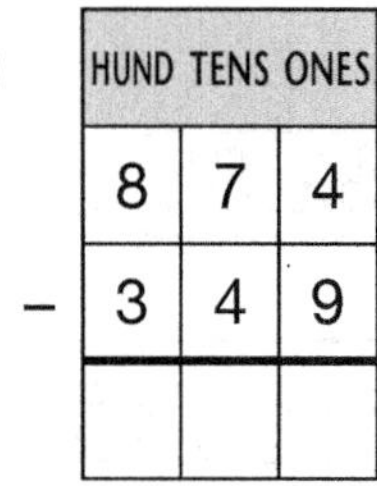

	HUND	TENS	ONES
	8	7	4
–	3	4	9

9 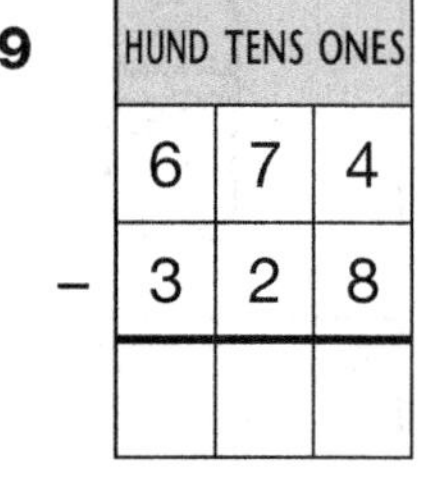

	HUND	TENS	ONES
	6	7	4
–	3	2	8

10 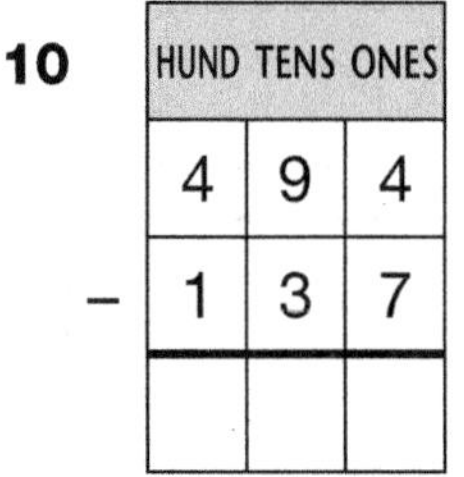

	HUND	TENS	ONES
	4	9	4
–	1	3	7

11 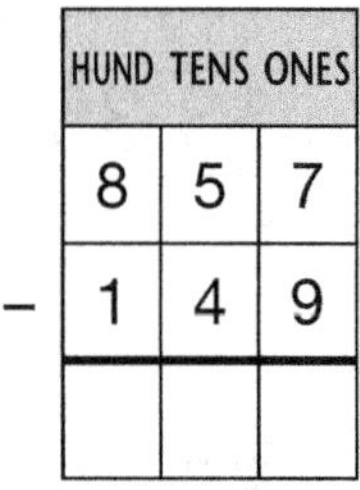

	HUND	TENS	ONES
	8	5	7
–	1	4	9

12

	HUND	TENS	ONES
	3	8	0
–	1	2	5

13 Sophie is saving to buy a $635 television. If she has saved $313, how much more does she need? 

Space Turns

Write the amount of each clockwise turn for the shapes ($\frac{1}{4}$, $\frac{1}{2}$, or $\frac{3}{4}$).

1 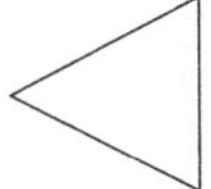  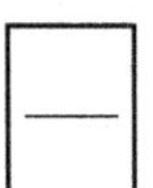

3 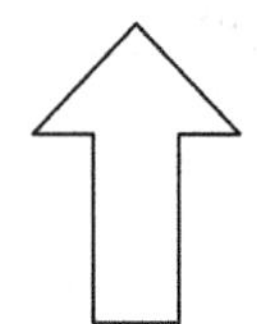 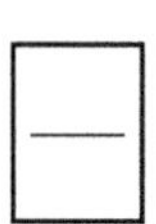

2 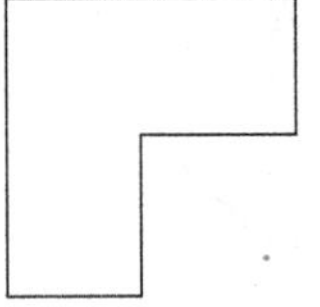  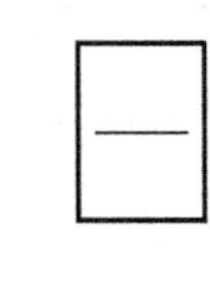 

4 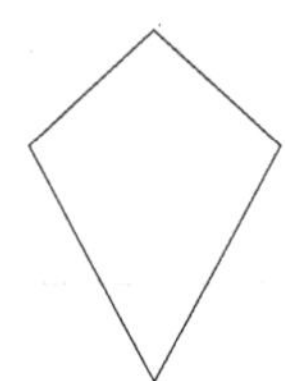 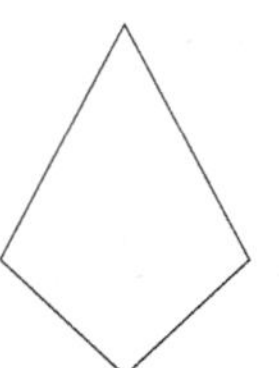 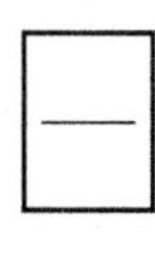

Number and Algebra

SET 3 Fractions of a quantity

Find the fractions by circling the shapes.

1 ● ● ● ● ● ● $\frac{1}{2}$ of 6

2 ▲ ▲ ▲ ▲ ▲ ▲ ▲ ▲ ▲ ▲ $\frac{1}{2}$ of 10

3 ■ ■ ■ ■ ■ ■ ■ ■ ■ ■ ■ ■ $\frac{1}{2}$ of 12

4 ● ● ● ● ● ● ● ● $\frac{1}{4}$ of 8

5 ▲ ▲ ▲ ▲ ▲ ▲ ▲ ▲ $\frac{1}{8}$ of 8

6 ■ ■ ■ ■ ■ $\frac{1}{5}$ of 5

Working Mathematically

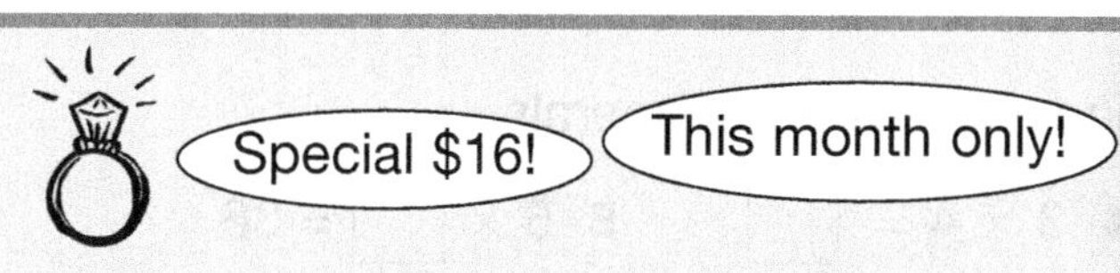

Four people are saving to buy this ring.

Calculate how much each person has saved if:

7 Chris has $\frac{1}{2}$ the amount needed. ____

8 Donna has $\frac{1}{4}$ the amount needed. ____

9 Tanya has $\frac{1}{8}$ the amount needed. ____

SET 4 Extension

1 What is the shortest month of the year?

2 How many days in December and September?

3 34, 37, 40, ☐, ☐

4 How many 10c coins in $9.30?

5 If the big hand is on the 12 and the small hand is on the 10, what time is it?

6 How many centimetres in $2\frac{1}{2}$ metres?

7 How many months in 2 years?

8 $5.70 take away 40c

9 185c = $ ☐

10 Share 48 among 6.

11 8 tens minus 8 ones

12 How many sides do 6 octagons have?

13 What is the time 5 hours after 6 pm?

14 50 cm is what fraction of 1 m?

15 2 kilograms = ☐ grams

16 What is the total number of sides for this group of shapes?

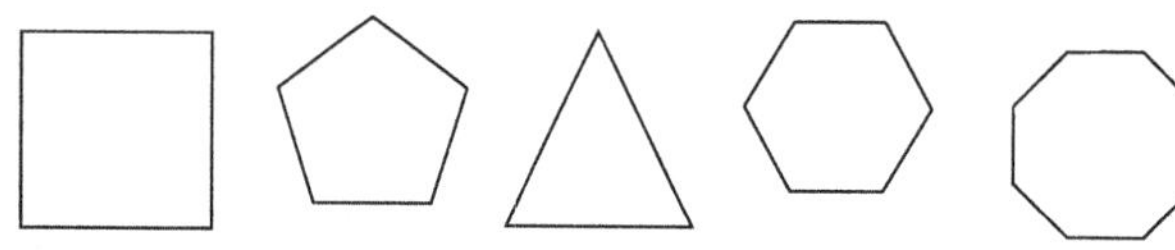

Measurement Time in minutes

Write the digital time that is 20 minutes later than the time shown on each clock face.

1 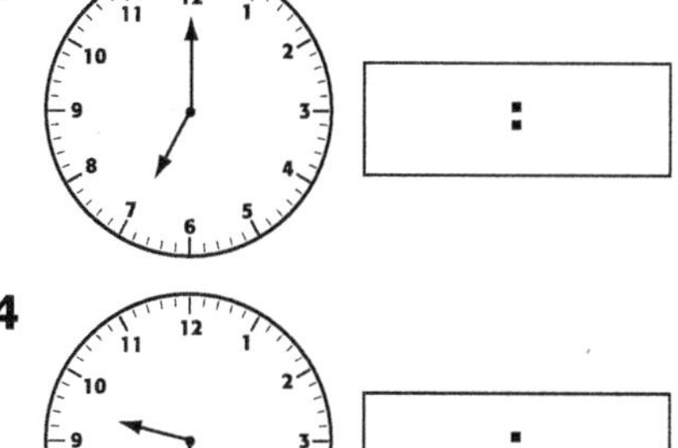 ☐ :

2 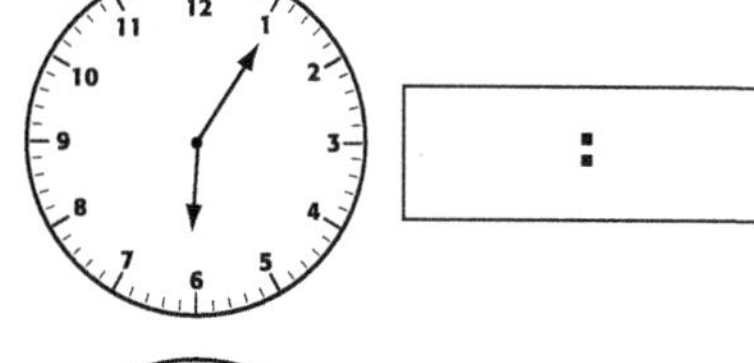 ☐ :

3 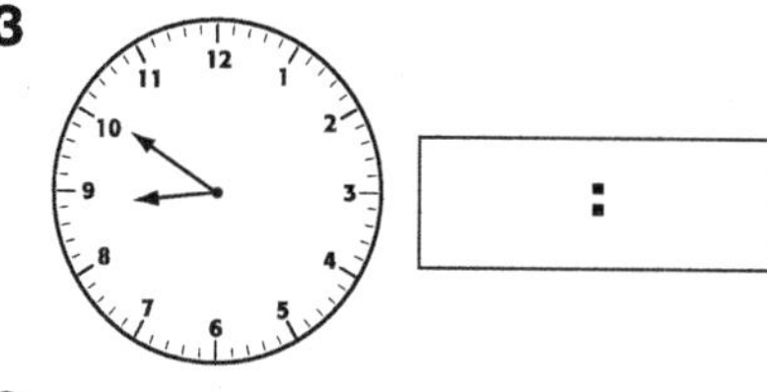 ☐ :

4 ☐ :

5  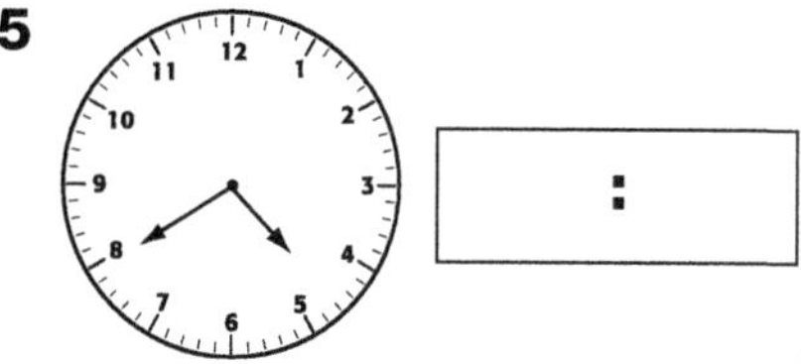 ☐ :

6 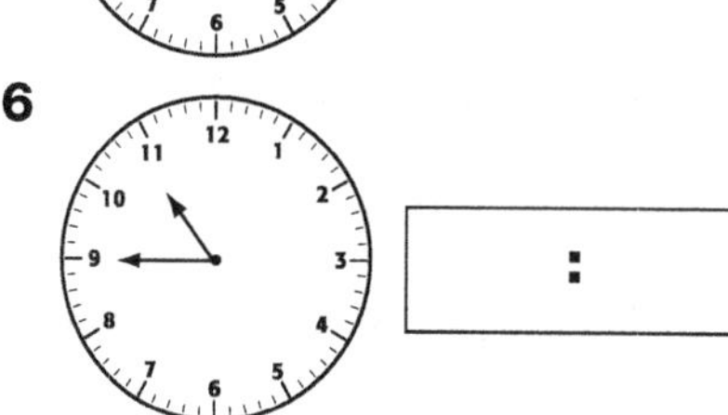 ☐ :

UNIT 23

Number and Algebra

SET 1 Basic

1 4 + 8

2 How many cents in $1?

3 $2 + $2 + $2

4 5c + ☐ = 10c

5 20c + 40c

6 15c – 10c

7 11 + 2 + 3

8 $5 × 0

9 $15 + $2

10 60c – 50c

11 Half of 20c

12 3 × 9

13 $1.20 + 5c

14 5 × 9

15

How many $5 notes are needed to make $40?

☐ notes

SET 2 Multiplication facts

1

× 5

5 2 1 7 3 8 6 4

2

× 3

5 2 1 7 3 8 6 4

3

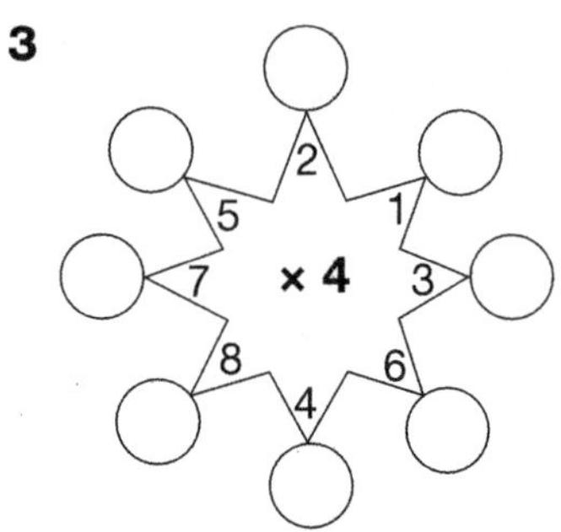

Find the missing numerals.

4 3 × 4 = ☐

5 6 × ☐ = 36

6 5 × ☐ = 40

7 9 × 10 = ☐

8 8 × ☐ = 0

9 8 × 6 = ☐

10 6 × 5 = ☐

11 7 × ☐ = 21

12 5 × 10 = ☐

13 4 × 6 = ☐

14 7 × ☐ = 35

15 3 × 8 = ☐

Space Grid references

Name the shapes for these grid references.

1 A4 = ☐

2 B3 = ☐

3 B1 = ☐

4 Draw a rectangle at C3.

5 Draw a triangle at A1.

6 What is the grid reference for the circle? ☐

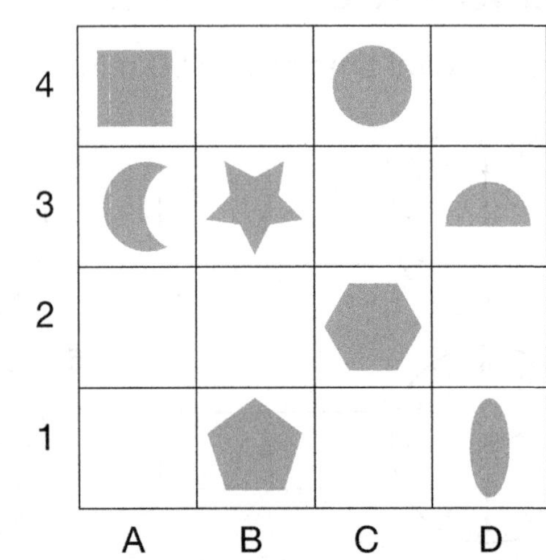

Number and Algebra

SET 3 Levelling and constant difference

Solve these additions using levelling.

1 36 + 46 becomes ☐ + ☐ = ☐

2 46 + 59 becomes ☐ + ☐ = ☐

3 29 + 38 becomes ☐ + ☐ = ☐

4 52 + 37 becomes ☐ + ☐ = ☐

5 64 + 88 becomes ☐ + ☐ = ☐

6 168 + 27 becomes ☐ + ☐ = ☐

Solve these substitutions using the constant difference strategy.

7 66 − 38 becomes ☐ + ☐ = ☐

8 72 − 46 becomes ☐ + ☐ = ☐

9 91 − 59 becomes ☐ + ☐ = ☐

10 85 − 37 becomes ☐ + ☐ = ☐

11 151 − 46 becomes ☐ + ☐ = ☐

12 182 − 69 becomes ☐ + ☐ = ☐

SET 4 Extension

1 $1.50 + $0.20 = ______

2 $1.45 + $0.40 = ______

3 $1.76 − $0.25 = ______

4 $4.99 − $0.66 = ______

5 Share $2 among 4 people.

6 $3.00 = ☐ c

7 The bus fare is 45c. How much will it cost for 6 trips?

8 How much did I spend if I bought an ice-block for 25c and a cheesestick for 65c?

Working Mathematically

9 Meat pie

$1.40

Apple pie

$1.30

Which pie did Kyra buy if she used a $5 note and received $3.70 change?

She bought the ______________.

Statistics and Probability Picture graphs

Favourite fruits

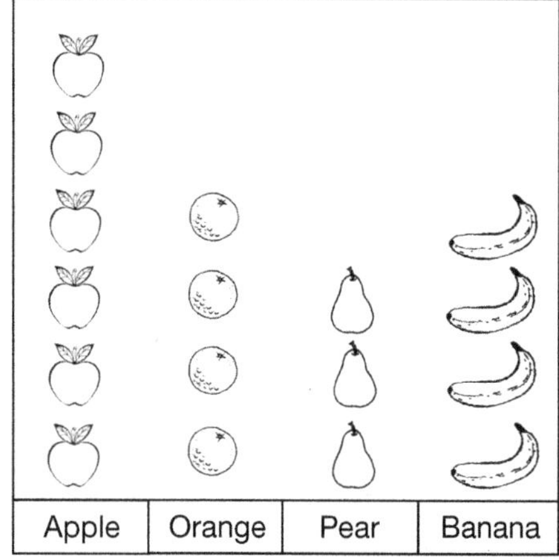

1 Which fruit was most popular?

2 Which fruit was least popular?

3 Which fruits were equally popular?

4 Which fruit scored 2 more than banana?

UNIT 24

Number and Algebra

SET 1 Basic

1 6 + 8

2 Sum of 6 and 5

3 Add 17 and 6.

4 18 – 5

5 19 – 6

6 20 – 7

7 Half of 50

8 ☐ + 7 = 19

9 10 × 5

10 5 × 6

11 Product of 7 and 5

12 Double 5 plus 10

13 Add 6 to 21.

14 What is the difference between 16 and 8?

15

Jody planted 5 beds of 6 flowers. How many flowers did she plant?

☐ flowers

SET 2 Compensation strategy

When adding 36 + 29, think 36 + 30 – 1 = 65

Use the strategy above to find answers to these equations.

	Equation	Compensation strategy	Answer
1	23 + 38	23 + 40 – 2	
2	32 + 29		
3	18 + 49		
4	25 + 39		
5	27 + 48		
6	46 + 19		
7	48 + 49		
8	53 + 28		

Find answers to the following.

9 39 + 19

10 41 + 38

11 16 + 79

12 24 + 59

13 39 + 69

Statistics and Probability Recording data

Create a graph from the data for the goals scored.

Name	Goals
Harry	卌 ||
Joe	||||
Sam	卌 卌
Peter	卌 ||||
Sue	|||
Lauren	卌 ||

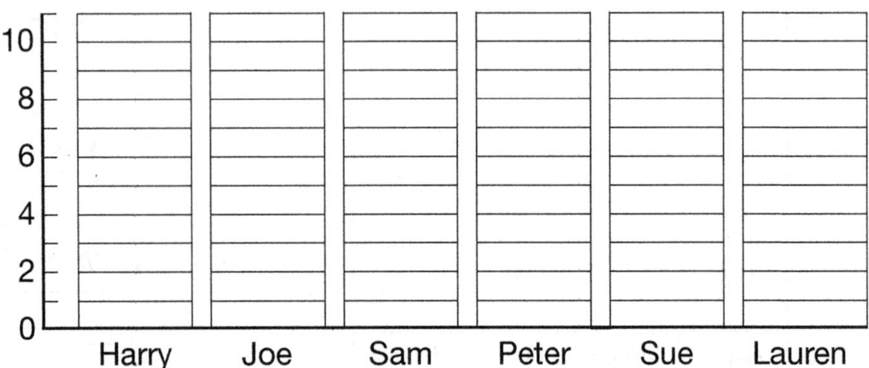

Number and Algebra

SET 3 Division problems

1 12 passengers were shared equally among 4 cars. How many in each car?

2 Mum baked 15 cakes which she shared among 5 children. How many cakes did each child get?

3 Samira shared 18 football cards among herself and 2 friends. How many did each child receive?

4 Two dozen eggs were shared among 4 families. How many did each family get?

5 25 sheep are shared among 5 paddocks. How many sheep are in each?

SET 4 Extension

1 How much are 5 books at $9 each?

2 8 hundreds + 7 tens

3 Estimate an answer to 133 + 42.

4 $1\frac{1}{2}$ kg of sugar at $2.00 per kilogram

5 Minutes in $4\frac{1}{2}$ hours

6 Write 272 in words.

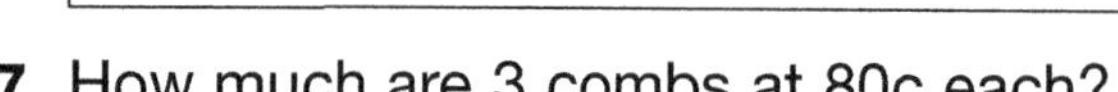

7 How much are 3 combs at 80c each?

8 Value of 3 in 7324

9 What is one-third of 90?

10 How many 50c cakes can I buy for $2.50?

Working Mathematically

11

Name 2 ways Dad could have spent $30.

	Quantity	$
T-shirt		
Tie		
Socks		
Total		

	Quantity	$
T-shirt		
Tie		
Socks		
Total		

Measurement Kilograms

Nick's dad has a mass that fits the following description:

- It is an odd number.
- It is between 70 kg and 100 kg.
- It has a 5 in the ones place.
- It is not 85 kg.
- It is 15 kg lighter than another mass shown here.

Circle Nick's dad.

65 kg	75 kg	85 kg	70 kg	90 kg

UNIT 25

Number and Algebra

SET 1 Basic

1 5 × 5

2 6 × 0

3 31 + 8

4 ☐ + 5 = 20

5 How many metres in a kilometre?

6 7 times 4

7 18 – 12

8 Half of 20

9 16 + 0 + 4

10 8 tens + 1 more ten

11 4 × 9

12 46 – 6

13 Two-quarters of 4

14 60 – 50

15 

SET 2 Addition problems

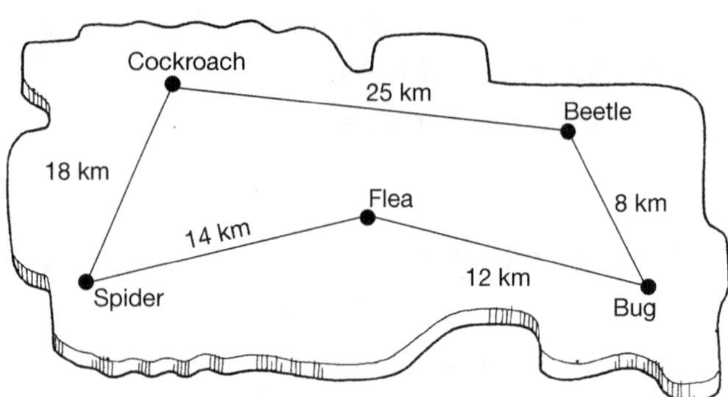

How many kilometres from:

1 Spider to Bug via Flea?

2 Spider to Beetle via Cockroach?

3 Cockroach to Bug via Beetle?

4 Cockroach to Flea via Spider?

5 Cockroach to Flea via Beetle and Bug?

Solve the problems.

6 Ben scored 35 and 46 in two tests. What was his total mark? ☐

7 Wali had $53 and Jing had $35. How much money did they have altogether? ☐

8 Suki has 2 dozen eggs and Katia has 27 eggs. How many eggs do they have altogether? ☐

Space Polygons

Colour the squares red, the rhombuses green, the trapeziums blue, and the rectangles yellow.

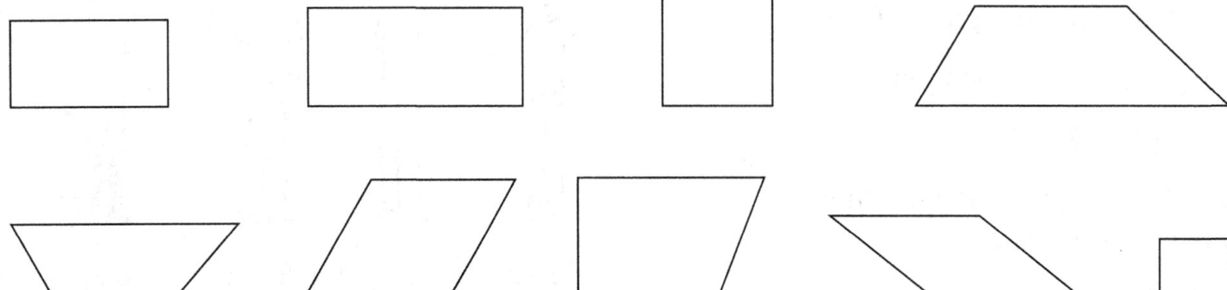

Number and Algebra

SET 3 Fractions on a number line

Draw a line to match the fractions to the number line.

1

$\frac{1}{4}$ $\frac{3}{4}$

0 — 1

2

$\frac{1}{5}$ $\frac{3}{5}$ $\frac{5}{5}$

0 — 1

3

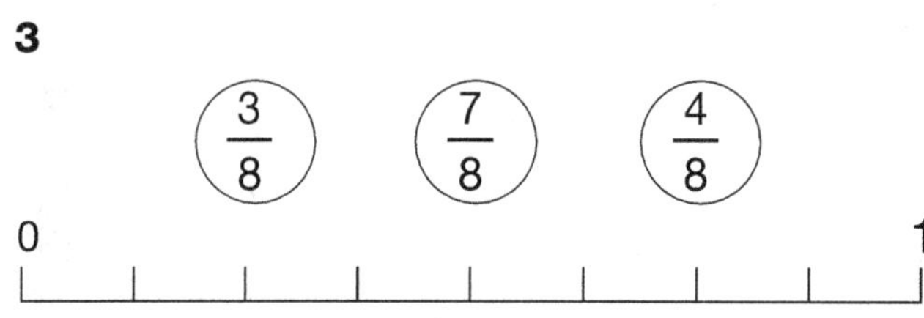

4

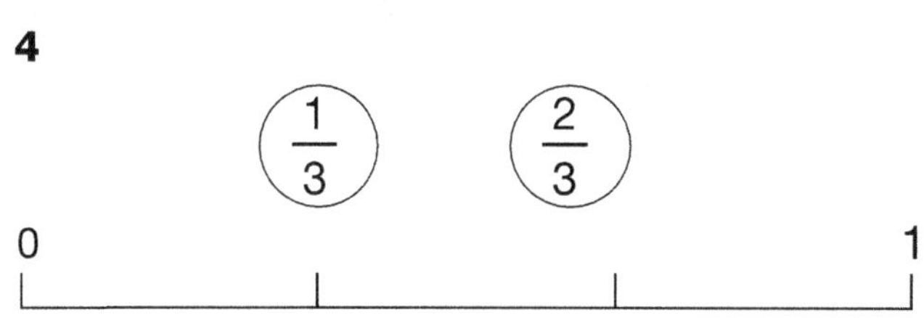

SET 4 Extension

1. How many tens in 360?
2. How many centimetres in $2\frac{1}{2}$ metres?
3. $\frac{1}{4}$ of 60
4. $8 \times \square = 40$
5. Product of 9 and 2
6. How many legs on 6 spiders?
7. Share 12 equally among 3 people.
8. 21 + 12 + 9
9. 17 tens + 19 ones
10. Share $100 among 5.
11. What is the difference between 73 and 8?
12. Amy is half her mother's age. How old is her mother if Amy is 24?
13. How many days are in a leap year?
14. If bracelets are $8.50 each, how much would 2 cost?
15. Which is longer, 1500 m or 2 km?
16. There are 193 pupils at St Mary's Primary School but 48 are on an excursion. How many are still at school?

Measurement The square metre

Colour the correct boxes to record the areas of the following.

Item	Smaller than 1 square metre	About 1 square metre	Larger than 1 square metre
Computer screen			
Beach towel			
Single bedsheet			
Garage floor			
Baby's blanket			
Tea towel			

Number and Algebra

SET 1 Basic

1 5 × 6

2 4 × 0

3 21 + 6

4 ☐ + 3 = 10

5 How many grams in a kilogram?

6 6 times 3

7 19 – 11

8 Half of 12

9 15 + 0 + 2

10 8 tens and 9 ones

11 6 × 4

12 35 – 4

13 A quarter of 8

14 31 – 21

15 How many days in April and May?

16 Which number sentence has 21 as its answer?

❍ 15 + 8 ❍ 30 – 8

❍ 3 × 6 ❍ 3 × 7

SET 2 3-digit subtraction

1

	HUND	TENS	ONES
	7	8	9
–	3	4	6

2

	HUND	TENS	ONES
	8	9	7
–	2	7	4

3

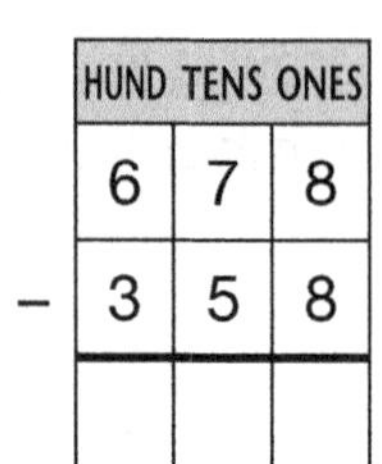

	HUND	TENS	ONES
	6	7	8
–	3	5	8

4

	HUND	TENS	ONES
	5	8	7
–	2	6	5

5

	HUND	TENS	ONES
	6	8	5
–	4	4	3

6

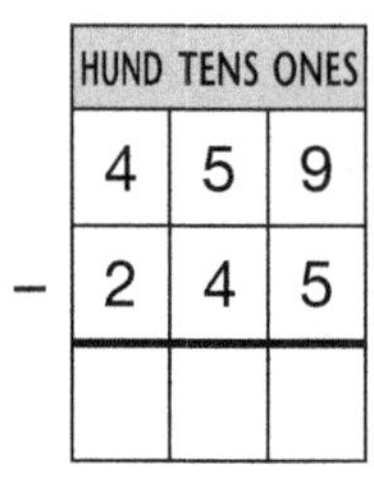

	HUND	TENS	ONES
	4	5	9
–	2	4	5

7 829 – 27

8 938 – 23

9 Subtract 26 from 738.

10 Subtract 37 from 267.

Working Mathematically

Use rounding to the nearest 100 to estimate the difference between the following numbers.

E.g. 798 – 302 becomes 800 – 300 = 500

		Estimate
11	600 – 198 =	
12	300 – 202 =	
13	800 – 395 =	
14	500 – 210 =	

Space Octagons/shapes

Join the dots to form the shapes then name them.

2• •3

1•——•4

3 2• •4 1•——•5

3• •4 2• •5 1•——•6

4• •5 3• •6 2• •7 1•——•8

1 ☐ 2 ☐ 3 ☐ 4 ☐

Number and Algebra

SET 3 4-digit numbers

Expand the numbers on the numeral expanders.

1 3674

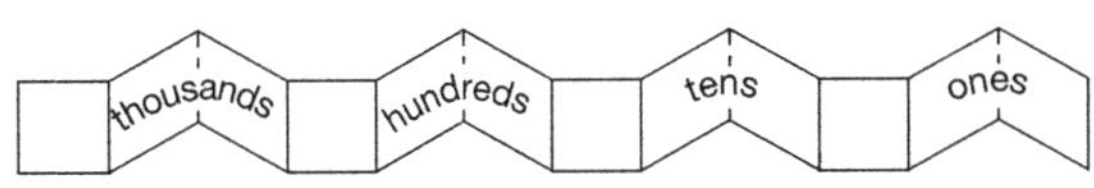

2 3842

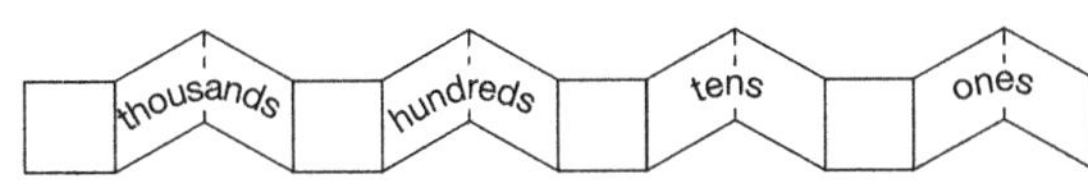

3 2357

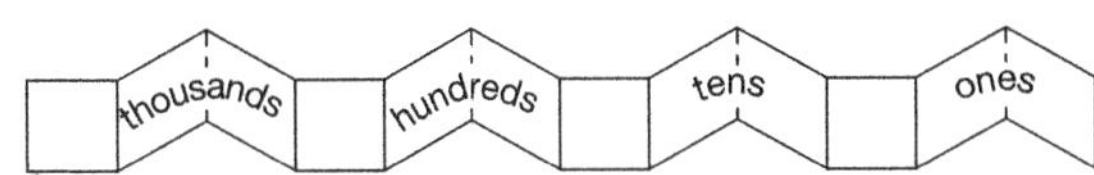

Record the place value of each bold number.

4 2**7**43 ______________

5 35**7**9 ______________

6 2**5**49 ______________

7 **3**567 ______________

8 570**9** ______________

9 **3**474 ______________

SET 4 Extension

1 How many tens in 450?

2 How many grams in a quarter of a kilogram?

3 $\frac{1}{4}$ of 16

4 $9 \times \square = 45$

5 Product of 6 and 3

6 6 lollies cost 30c, how much for 4?

7 Share 48 among 4.

8 16 + 4 + 160

9 Which is larger: $\frac{1}{4}$ or $\frac{1}{10}$?

10 7 tens + 21 ones

11 What is the difference between 642 and 9?

12 Xia is 9 and Chi is twice that age. How old is Chi?

13 How many days are there in a normal year?

14 If each watch costs $16, how much would 4 watches cost?

15 Round 2431 to the nearest 100.

Measurement Cubic centimetres

Put a cross on all the objects made from 12 cubic centimetres.

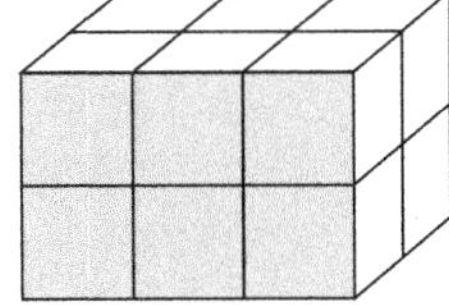

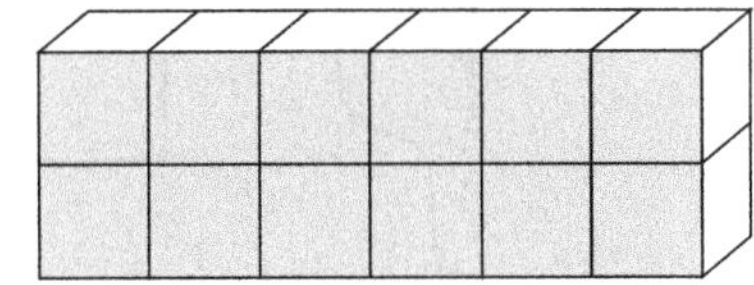

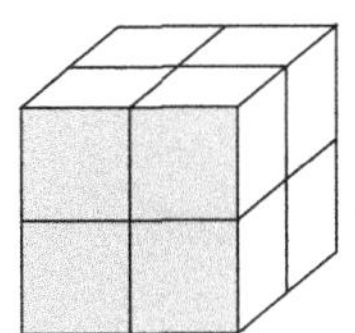

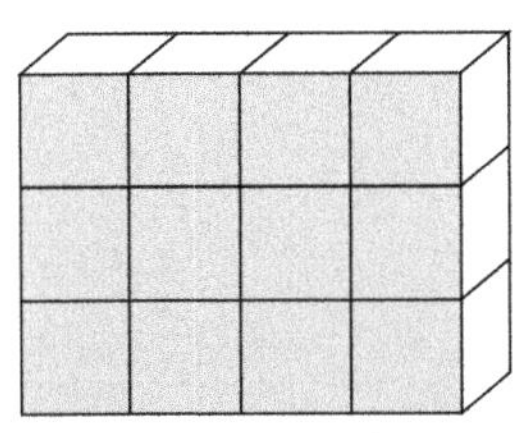

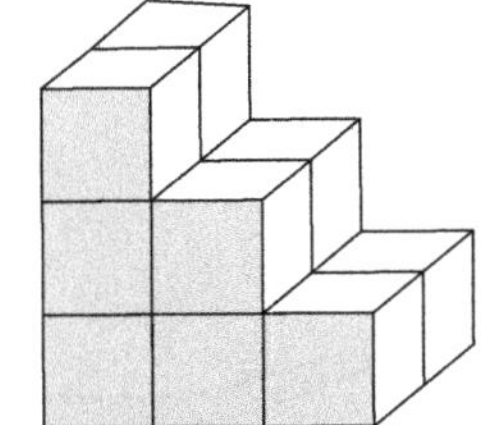

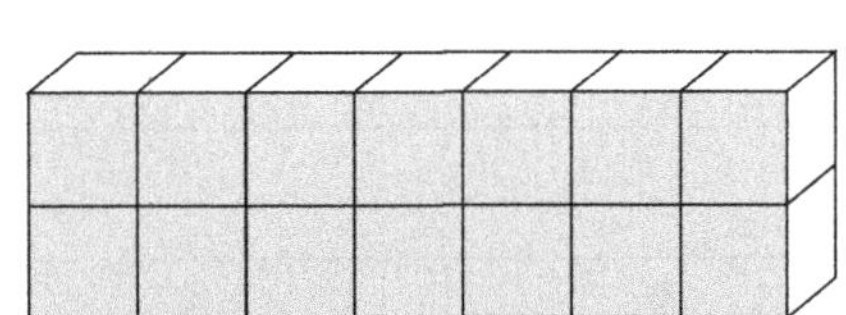

UNIT
27

Number and Algebra

SET 1 Basic

1 40 + 7

2 11 – 8

3 8 + 9 + 2

4 Half of 20c

5 Product of 5 and 8

6 Is 25 an odd number?

7 10 × 2

8 8 × 4

9 150 – 1

10 26 + 13

11 Product of 3 and 4

12 Is December a summer month?

13 21 – 10

14 4 lots of 6

15 

SET 2 Multiplication facts

Complete the tables.

	×	0	10	2	3	7	6	8
1	3							
2	4							
3	10							
4	5							

5 How much are 7 books at $5 each?

6 How much are 7 pens at $6 each?

7 How much are 6 calculators at $4 each?

Complete the sets of multiples.

8	5	10	15					
9	4	8	12					
10	3	6	9					
11	9	18	27					
12	6	12	18					

13 True or false? 24 is a multiple of 2, 3, 4, 6, 8 and 12.

Space Giving directions

Complete the instructions to describe the path drawn on the map.

1 Start at ■. Travel east 4 spaces.

2 Travel north 3 spaces.

3 ____________________

4 ____________________

5 ____________________

6 ____________________

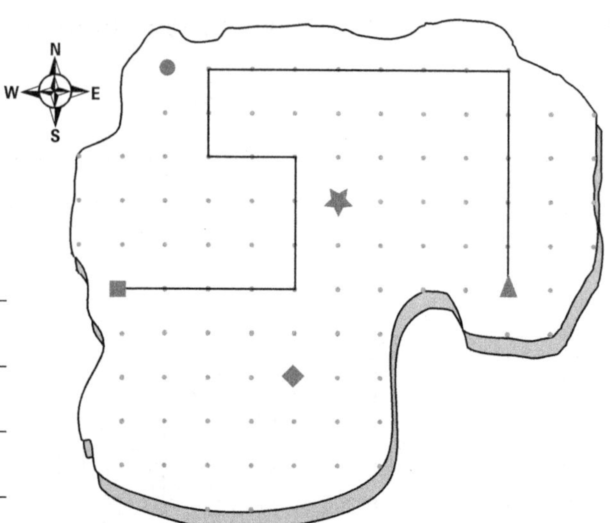

Number and Algebra

SET 3 Division

Use Base 10 materials to answer these questions.

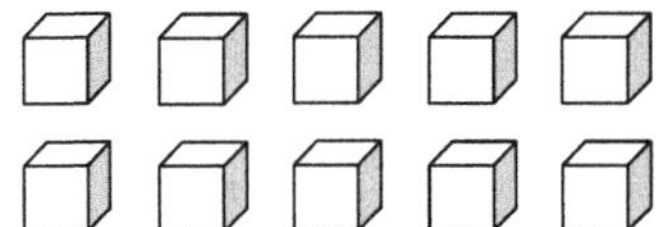

1 How many groups of 5? ☐

2 How many groups of 2? ☐

Use the triangles to solve these divisions.

△ △ △ △ △ △ △ △ △ △
△ △ △ △ △ △ △ △ △ △

3 $20 \div 2 =$ ☐

4 $20 \div 5 =$ ☐

5 $20 \div 4 =$ ☐

6 $20 \div 10 =$ ☐

Answer the questions

7 $8 \div 2 =$ ☐

8 $15 \div 5 =$ ☐

9 $15 \div 3 =$ ☐

10 $16 \div 4 =$ ☐

11 $50 \div 10 =$ ☐

SET 4 Extension

1 $5.63 + 0.17 = ____

2 188 km + 14 km = ____

3 46 hours − 26 hours = ____

4 295 cm − 124 cm = ____

5 How much are 3 kg of potatoes at $2.50 kg?

6 Peter has $27 for his parents' present and his sister has $92. How much do they have altogether?

Working Mathematically

7 Place the numbers 6, 2, 1 and 7 into the shapes so that each line adds to 13.

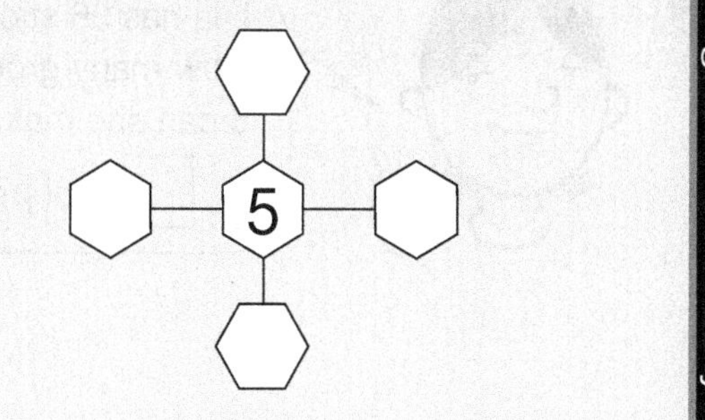

Statistics and Probability Chance

Colour the marbles in the bag to make the following statements true.

1 A red marble is most likely to be drawn out of the bag.

2 A green marble is least likely to be drawn out of the bag.

3 Blue will never be drawn out of the bag.

UNIT 28

Number and Algebra

SET 1 Basic

1 Sum of 6 and 7

2 3 + ☐ = 9

3 13 + ☐ = 20

4 59 + 3

5 48 – 5

6 99 – 9

7 2 × 6

8 4 × 6

9 ☐ × 6 = 18

10 5 × 6

11 7 × 6

12 8 × 6

13 How many sixes make 12?

14 How many sixes make 24?

15 

Lila has 36 stickers. How many groups of 6 can she make?

☐ groups of 6

SET 2 Checking work

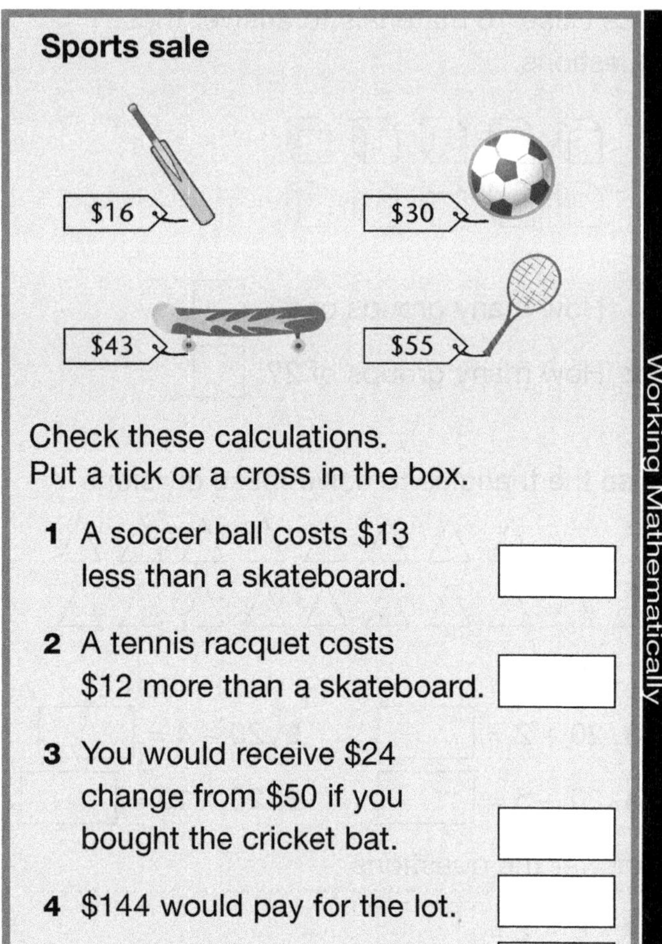

Check these calculations.
Put a tick or a cross in the box.

1 A soccer ball costs $13 less than a skateboard. ☐

2 A tennis racquet costs $12 more than a skateboard. ☐

3 You would receive $24 change from $50 if you bought the cricket bat. ☐

4 $144 would pay for the lot. ☐

5 3 skateboards cost $139. ☐

Space Pyramids

Colour the object that fits the set of faces.

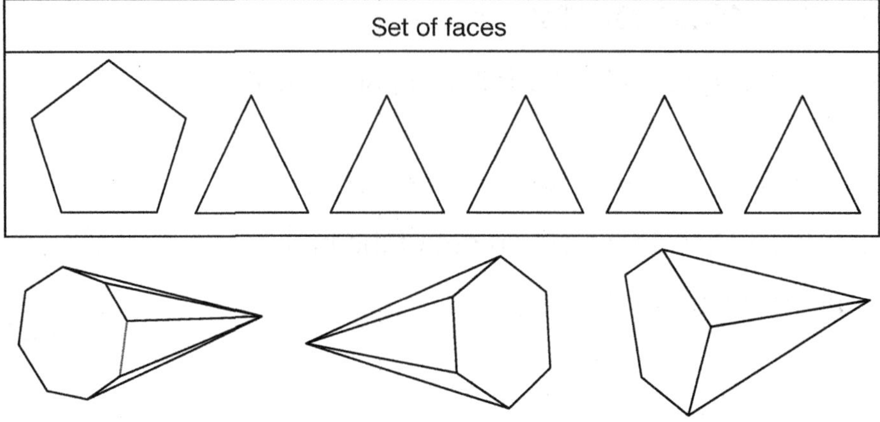

Number and Algebra

SET 3 Commutative law

Write answers for the pairs of additions.

1	8 + 7 =	7 + 8 =
2	9 + 6 =	6 + 9 =
3	12 + 13 =	13 + 12 =
4	12 + 9 =	9 + 12 =
5	20 + 15 =	15 + 20 =
6	15 + 8 =	8 + 15 =
7	25 + 13 =	13 + 25 =
8	40 + 29 =	29 + 40 =

9 Does it matter in what order numbers are added?

Complete each multiplication pair.

10	5 × 2 =	2 × 5 =
11	6 × 2 =	2 × 6 =
12	3 × 5 =	5 × 3 =
13	4 × 5 =	5 × 4 =
14	7 × 5 =	5 × 7 =
15	6 × 4 =	4 × 6 =

16 Does it matter in what order numbers are multiplied?

SET 4 Extension

1 ☐ × 20 = 80

2 Jim spent 7c and 13c. If he had 50c, how much has he left?

3 61, 65, 69, ☐ , ☐

4 Three cakes cost \$9. How much for 4?

5 Is a prism a 2D shape?

6 How many 50c coins make \$12.00?

7 182 – 5

8 Write the odd numbers between 82 and 90.

9 How many centimetres in $7\frac{1}{2}$ m?

10 Write the number 9 less than 480.

11 How many angles has a square?

12 286c = \$ ☐

13 What are the factors of 16?

14 Double 4 then double again.

15 If eggs cost \$1.20 a dozen, how much for $4\frac{1}{2}$ dozen?

16 Share 33 among 5.

Measurement Kilograms

Number these items from lightest to heaviest.

☐

☐

☐

☐

☐

UNIT 29

Number and Algebra

SET 1 Basic

1 9 + ☐ = 16

2 20c – 15c

3 3 × 3

4 16 + 3 + 5

5 Half of 24

6 17 – 6

7 29 + 7

8 100 – 30

9 5 × 10

10 Sum of 21 and 8

11 5c + 5c + 5c

12 7 × 4

13 $5 = ☐ cents

14 20 minus 13

15

Michael had 6 lots of 10 popsticks. How many popsticks did he have altogether?

☐ popsticks

SET 2 3-digit addition

1

	HUND	TENS	ONES
	3	2	4
+	4	1	6

2

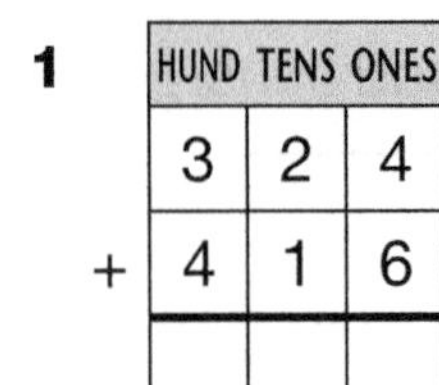

	HUND	TENS	ONES
	5	6	7
+	2	2	5

3

	HUND	TENS	ONES
	3	4	8
+	4	3	7

4

	HUND	TENS	ONES
	1	5	6
+	2	6	6

5

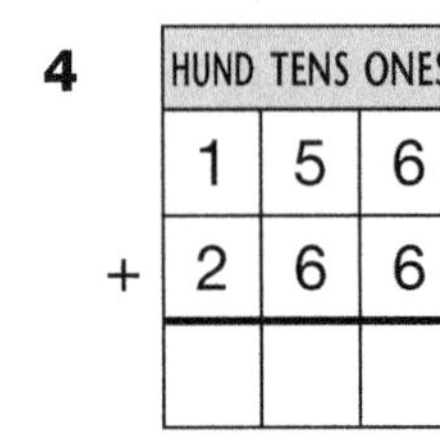

	HUND	TENS	ONES
	3	7	7
+	4	7	7

6

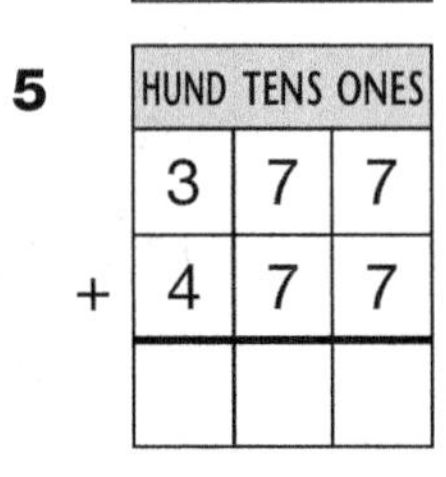

	HUND	TENS	ONES
	3	4	7
+	2	8	4

Calculator

Place the jewellery in order of value from cheapest to dearest. Work out the total value of the jewellery.

$3499

$5499

$290

$4999

	Item	Value
7		
8		
9		
10		
	Total	

Space Parallelograms

Shade all parallelograms.

Tick all squares.

Put a cross on all rectangles.

Circle all rhombuses.

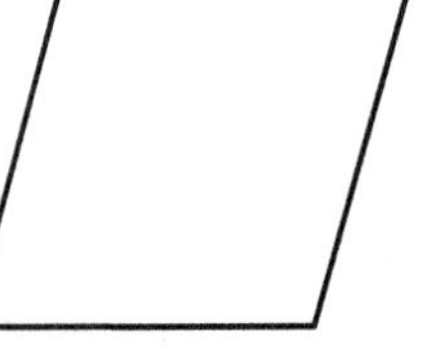

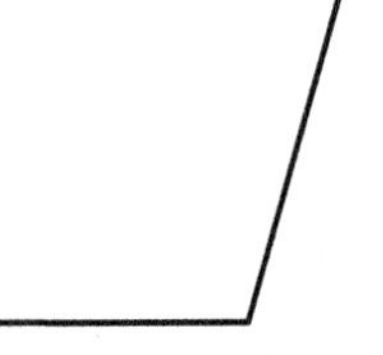

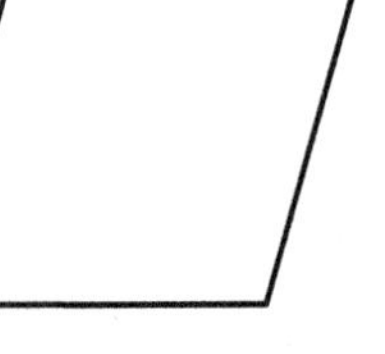

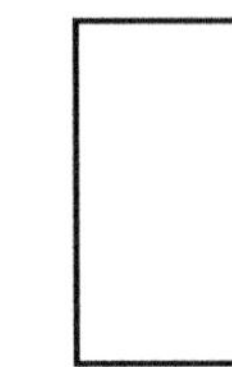

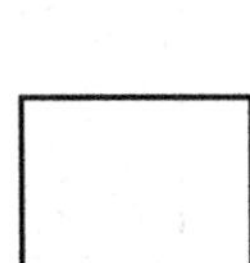

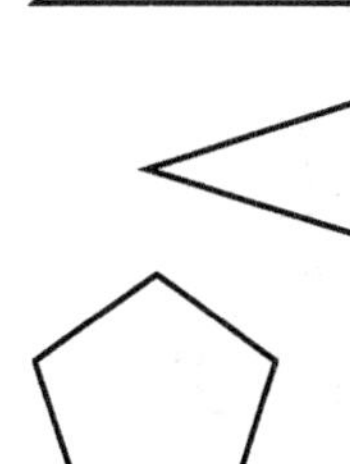

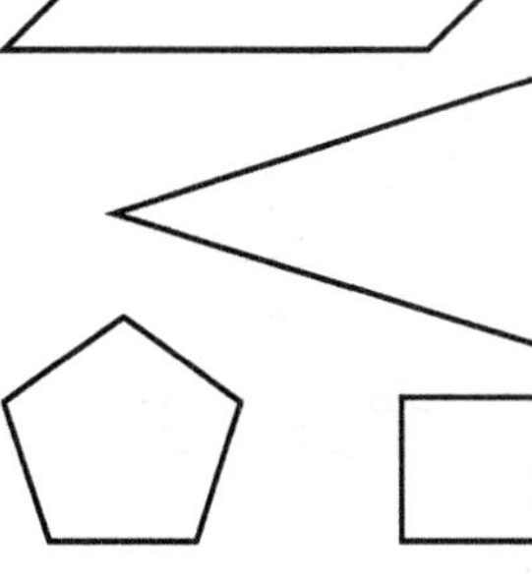

Number and Algebra

SET 3 Division

Use the pizzas to help you solve the divisions.

1 16 ÷ 8 = ☐

2 16 ÷ 4 = ☐

3 16 ÷ 2 = ☐

Use the apples to help you solve the problems.

4 Bejo shared 18 apples among 2 boys and himself. How many did each boy receive?

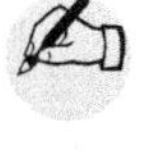

5 Six girls shared 18 apples. How many did each girl receive?

SET 4 Extension

1
$$\begin{array}{r} 466 \\ 277 \\ +\ \ 43 \\ \hline \end{array}$$

2
$$\begin{array}{r} 324 \\ 28 \\ +\ 639 \\ \hline \end{array}$$

3
$$\begin{array}{r} 424 \\ -\ 318 \\ \hline \end{array}$$

4
$$\begin{array}{r} 645 \\ -\ 328 \\ \hline \end{array}$$

5 Share 27 among 6.

6 9 less than 256

7 325, 350, 375, ☐, ☐

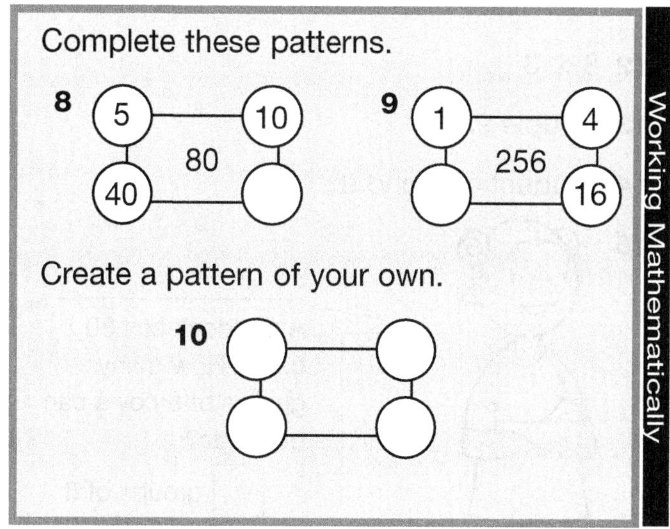

Statistics and Probability Chance experiments

Here are the results of spinning this spinner 20 times.

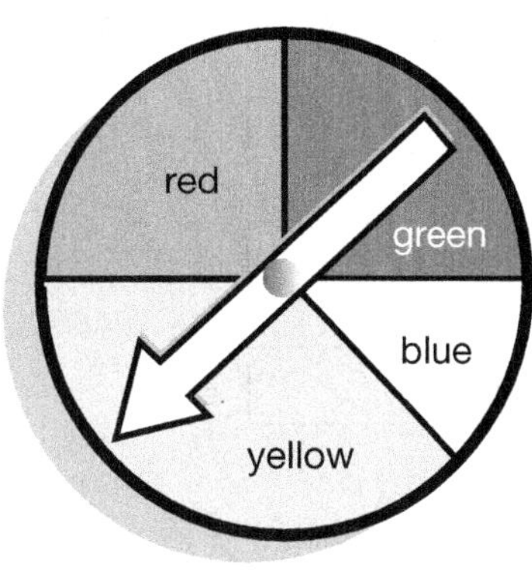

Red	𝍸
Green	||||
Blue	||
Yellow	𝍸 ||||

1 What colour occurred most often?

2 Do you think the spinner is more likely to land on red than on blue? ________

3 a Which colour has the least chance of occurring? ________

b Do the results show this?

UNIT 30

Number and Algebra

SET 1 Basic

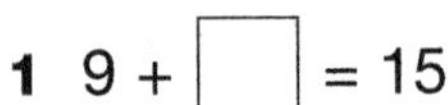

1 9 + ☐ = 15

2 \$20 – \$7

3 3 + 7 + 3 + 7

4 22 + 8

5 6×3

6 Sum of 24 and 9

7 4×4

8 Six lots of 6

9 27 – 6

10 9×5

11 Four rows of 8

12 8×3

13 Double 9.

14 Product of 4 and 9

15

A paddock has 30 cows. How many groups of 6 cows can be made?

☐ groups of 6

SET 2 Addition to 9999

Add these numbers.

1

	THOU	HUND	TENS	ONES
	4	1	7	3
+	3	2	2	5

2

	THOU	HUND	TENS	ONES
	3	4	3	6
+	2	5	5	6

3

	THOU	HUND	TENS	ONES
	5	4	0	7
+	2	4	6	8

4

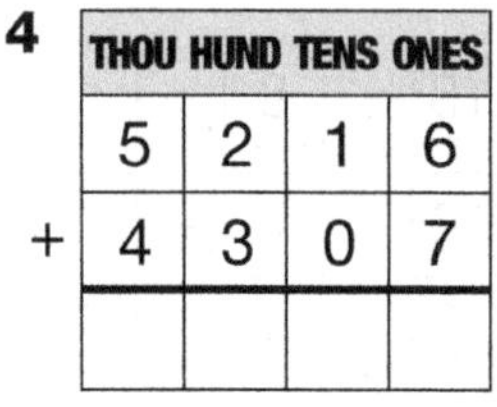

	THOU	HUND	TENS	ONES
	5	2	1	6
+	4	3	0	7

5

	THOU	HUND	TENS	ONES
	3	4	5	7
+	2	3	2	5

6

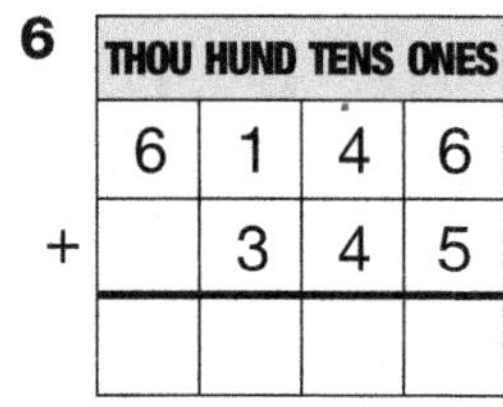

	THOU	HUND	TENS	ONES
	6	1	4	6
+		3	4	5

7 Jack had \$1342 and was given \$500 more. How much does he have now?

8 Gina had \$3356 and was given another \$5000. How much does she have now?

Number and Algebra Levelling in subtraction/estimation

Solve these subtractions using levelling.

1 52 – 16 becomes 56 – 20 = ☐

2 46 – 28 becomes ☐ – ☐ = ☐

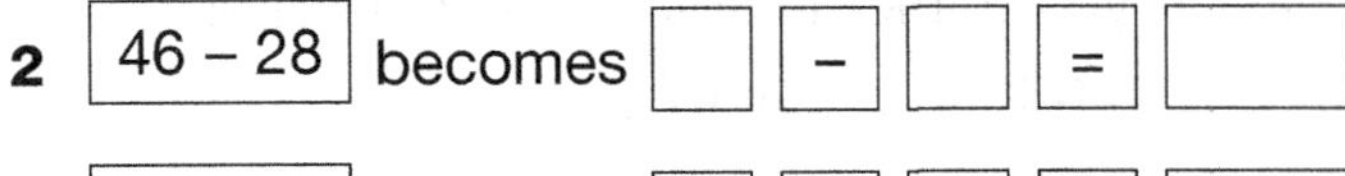

3 32 – 14 becomes ☐ – ☐ = ☐

4 163 – 57 becomes ☐ – ☐ = ☐

5 194 – 77 becomes ☐ – ☐ = ☐

Round each number to ten in the subtraction number sentences to estimate an answer.

6 64 – 29 is about ☐

7 72 – 21 is about ☐

8 86 – 34 is about ☐

Number and Algebra

SET 3 Complementary fractional parts

Write the shaded fraction and the complement for each model below.

1 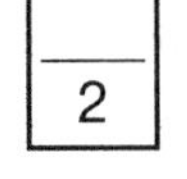 and $\frac{\square}{2}$

2 $\frac{\square}{\square}$ and $\frac{\square}{\square}$

3 $\frac{\square}{\square}$ and $\frac{\square}{\square}$

4 $\frac{\square}{\square}$ and $\frac{\square}{\square}$

5 $\frac{\square}{\square}$ and $\frac{\square}{\square}$

6 $\frac{\square}{\square}$ and $\frac{\square}{\square}$

SET 4 Extension

1 What is the value of 4 in 743?

2 What is the value of 6 in 637?

3 What digit represents thousands in 3504?

4 How many hours from noon to 6 pm?

5 407 = ☐ hund + ☐ tens + ☐ ones

6 Write the largest number you can using 1, 3, 2.

7 How many days in spring?

8 What 2 coins make 70c?

9 7 stickers at 15c each.

10 How many halves in $3\frac{1}{2}$?

Working Mathematically

Find pairs of numbers that produce these sums and products.

	Number	Number	Sum	Product
11	3	2	5	6
12			7	12
13			15	50
14			13	36
15			13	40

Space Quadrilaterals

Tick the quadrilaterals.

 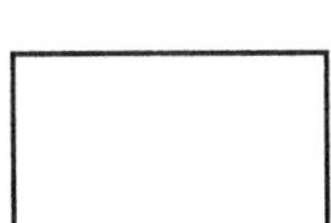 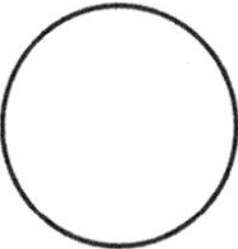 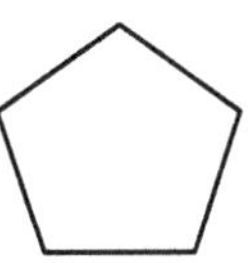 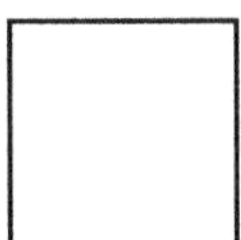 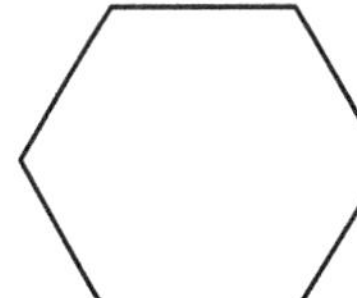 

Explain what all quadrilaterals have in common.

____

____

____

Number and Algebra

SET 1 Basic

1 ☐ + 7 = 15

2 3 + 19

3 2 + 38

4 27 − ☐ = 21

5 23 − 5

6 Product of 6 and 7

7 4 × 6

8 5 × 9

9 6 × 4

10 Sum of 17 and 8

11 9 tens and 6 ones

12 How many tens in 90?

13 Half of 20 then add 7

14 How many tens in 67?

15

SET 2 Writing and ordering numbers

1 Write the number for 10 hundreds.

2 Write the number for 15 tens.

3 Write the number for 4 thousands and 6 ones.

4 What is the place value of 7 in 3476?

5 What is the place value of 6 in 6423?

6 Order these numbers from smallest to largest: 1234, 2341, 1341. ______

7 Write the largest number you can using 5, 6 and 3. ______

8 Write the smallest number you can using 9, 4 and 1. ______

9 Write the number ten more than 4501. ______

Working Mathematically

10 Enter the number 6549 on your calculator.

Use subtraction to change the 4 to zero.

What buttons did you press? ______

______

What is your new number? ______

Space Nets

Draw a line to match the box to its net.

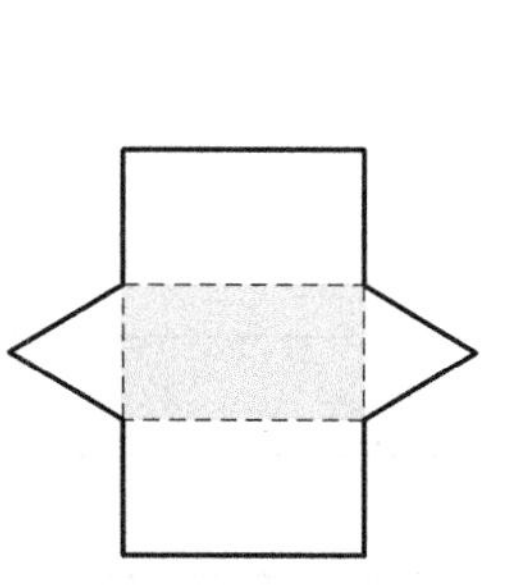

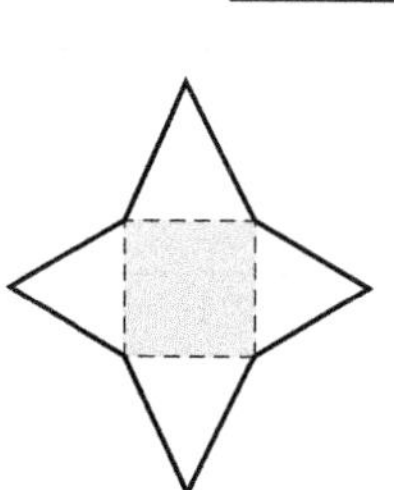

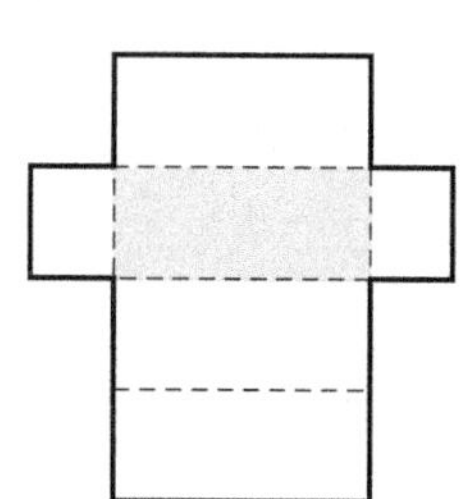

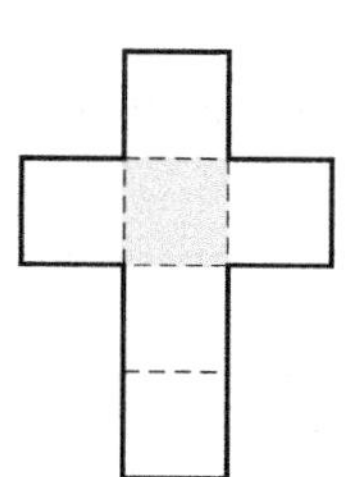

Number and Algebra

SET 3 Number patterns

Follow the rules to make number patterns.

1

Add 3							
7							

2

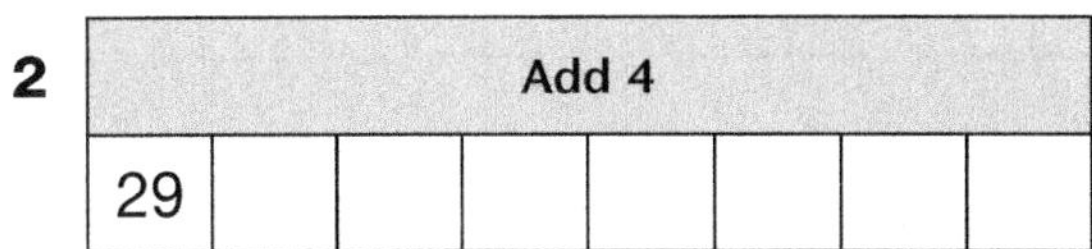

Add 4							
29							

3

Add 5							
61							

4

Add 6							
17							

Use your own strategy to solve the following patterns.

5	35	40	45	50				
6	80	75	70	65				
7	93	88	83	78				

SET 4 Extension

1 Write the numeral for 7 hundreds + 3 tens + 4 ones.

2 Write the numeral for 7 hundreds and 7 ones.

3 212, 217, 222, ☐ , 232

4 How many 5c lollies can be bought for $1.40?

5 42 + 18 + 43

6 907c = $ ☐

7 Share 36 among 5.

8 7 more than 25

9 How much is $4\frac{1}{2}$ kg of meat at $6 per kg?

10 What is the value of 6 in 764?

11 Are five 50c coins worth more than $2?

12 Subtract 70c from $8.00.

13 What is the value of 70c and 80c?

14 Add 5 tens to 341.

15 Which is heavier, $1\frac{1}{2}$ kg of feathers or 1 kg of metal?

16 Luca saved $376 in November and $561 in December. Does he have enough to buy a bike worth $899?

Measurement The gram

List 4 items that have their mass measured in grams.

Item	Mass

Number and Algebra

SET 1 Basic

1 Add 4 and 34.

2 14 + 7

3 3 + 4 + ☐ = 20

4 Product of 3 and 6

5 3 × 5 + ☐ = 20

6 Multiply 5 and 3.

7 37 – 5

8 Subtract 6 from 25.

9 47 minus 6

10 Sum of 6 and 27

11 Double 21.

12 3 + 3 + ☐ = 16

13 18 plus 7

14 Half of 42

15

SET 2 Number patterns

Continue the patterns.

1	13	15	17					
2	5	15	25					
3	90	85	80					
4	$\frac{10}{10}$	$\frac{9}{10}$	$\frac{8}{10}$					
5	48	42		30			12	
6	6	12		24				48

Working Mathematically

Create your own patterns.

7								
8								
9								

Space Describing 2D shapes

Below each shape record its name and the number of vertices it has.

Name					
Vertices					

Number and Algebra

SET 3 Bar models/subtraction

Use the bar models to help you complete the subtractions.

1

68	
	40

68 – 40 = ☐

2

82	
	45

82 – 45 = ☐

3

65	
	39

65 – 39 = ☐

4

74	
	19

74 – 19 = ☐

5

56	
	17

56 – 17 = ☐

6

43	
	28

43 – 28 = ☐

You can write
26 – 14 = 12
26 – 12 = 14

26	
14	12

SET 4 Extension

1 2 weeks + 6 days = ☐ days

2 What is the value of 8 in 8076?

3 Double 260.

4 How many times can 60 be subtracted from 240?

5 How much will 4 cakes cost at 75c each?

6 What is the value of 3 in 83 921?

7 400 minus 3

8 What number is 7 more than 156?

9 $(3 \times 100) + (2 \times 10) + (6 \times 1)$

10 378c = $ ☐

11 How much will $4\frac{1}{2}$ kg of meat cost at $4 kg?

12 Round 3579 to the nearest 10.

13 How many minutes from 9 am to 2 pm?

14 If lollies are 4 for 20c, how much would 9 cost?

15 How many hours from 3:30 am to 6:30 am?

16 Which number is 100 more than 375?

❍ 100 ❍ 400

❍ 475 ❍ 1375

Measurement Cubic centimetres

Record the volume of each object made from centicubes in cubic centimetres.

1

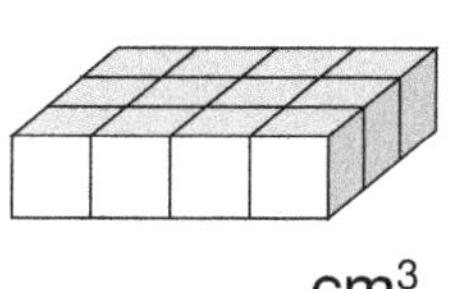

_________ cm^3

2

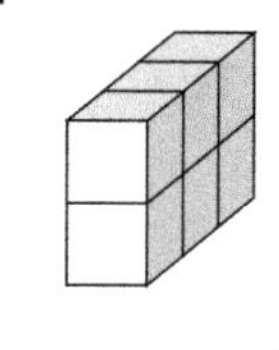

_________ cm^3

3

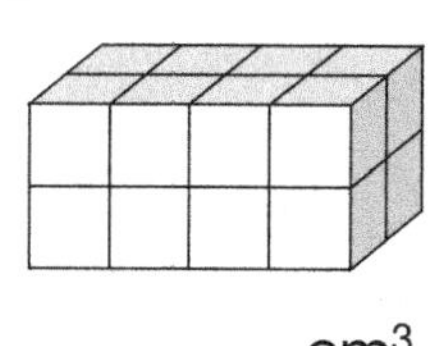

_________ cm^3

4

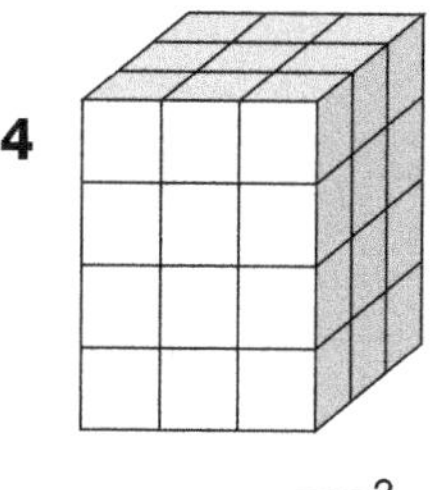

_________ cm^3

Number and Algebra

SET 1 Basic

1 19 + 3

2 10 × 9c

3 9 lots of 4

4 ☐ tens + ☐ ones = 57

5 6 × 3

6 Double 8.

7 How many vertices are in a triangle?

8 1994 + 5

9 Write 152 in words.

☐

10 300 + 40 + 7

11 $1.72 = ☐ cents

12 5 × 10c

13 Four squared

14 4 × 9

15 Write the smallest number you can using 7, 3, 1, 5. ☐

SET 2 Subtraction to 9999

1

	THOU	HUND	TENS	ONES
	6	7	4	8
−	3	4	3	2

2

	THOU	HUND	TENS	ONES
	3	4	5	2
−	1	2	2	6

3

	THOU	HUND	TENS	ONES
	8	6	7	4
−	3	5	3	6

4

	THOU	HUND	TENS	ONES
	2	7	3	8
−	1	3	2	9

5

	THOU	HUND	TENS	ONES
	3	7	8	4
−	1	5	3	6

6

	THOU	HUND	TENS	ONES
	8	7	6	2
−	6	5	2	7

7 Kiaan wants to buy a new television. If it costs $1723 and he has already saved $346, how much more does he need?

Number and Algebra Non-standard partitioning

Complete the numeral expanders to show how the numbers can be partitioned in different ways.

1

4 hundreds 1 tens 8 ones

☐ ☐ tens ☐ ones

2

5 hundreds 7 tens 9 ones

☐ ☐ tens ☐ ones

☐ ☐ ☐ ones

How many tens in each number?

3 165

4 248

5 337

6 581

How many hundreds in each number?

7 1875

8 2681

9 5811

10 4263

Number and Algebra

SET 3 Associative property

1

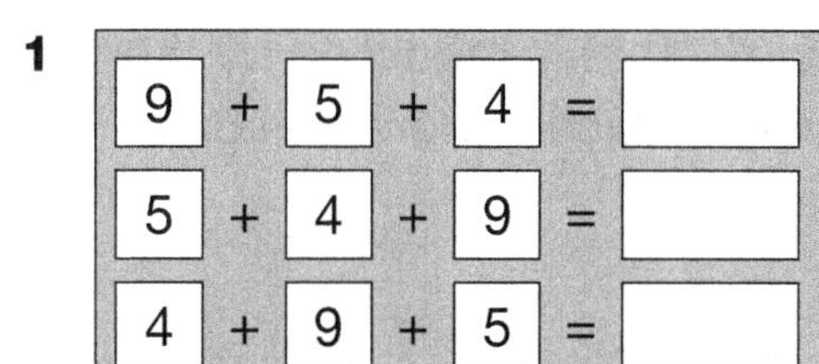

9 + 5 + 4 = ☐
5 + 4 + 9 = ☐
4 + 9 + 5 = ☐

2

6 + 7 + 5 = ☐
7 + 5 + 6 = ☐
5 + 6 + 7 = ☐

3

2 × 3 × 6 = ☐
3 × 6 × 2 = ☐
6 × 2 × 3 = ☐

4

1 × 5 × 2 = ☐
5 × 2 × 1 = ☐
2 × 1 × 5 = ☐

5 Does it matter in which order numbers are added?

6 Does it matter in which order numbers are multiplied?

SET 4 Extension

1 What is the sum of 4, 5 and 8?

2 7 kilograms = ☐ grams

3 Which is larger, 4×5 or 20×0?

4 Which is larger, 0.3 or $\frac{4}{10}$?

5 What number is 9 more than 1132?

6 How many minutes in $2\frac{1}{2}$ hours?

7 Ten times zero

8 How many grams in $1\frac{1}{4}$ kg?

9 Add 2 tens to 438.

10 What is the product of 4 and 8?

11 Share 21 among 5.

12 How many 50c coins are needed to make $14?

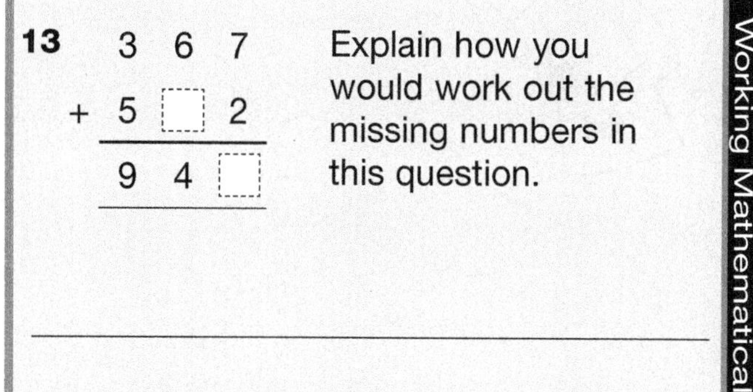

13

$$\begin{array}{r} 3\ 6\ 7 \\ +\ 5\ \square\ 2 \\ \hline 9\ 4\ \square \end{array}$$

Explain how you would work out the missing numbers in this question.

Working Mathematically

Measurement Millilitres

Order these containers from smallest to greatest capacity by numbering them 1 to 6.

1

Milk
1 Litre

☐

2

☐

3

☐

4

☐

5

☐

6

☐

Number and Algebra

SET 1 Basic

1 $8 $\times$ 10

2 1999 + 5

3 Sum of 13 and 27

4 Double 15.

5 How many tens in 80?

6 ☐ + 8 = 37

7 5^2

8 Double 8 and double again.

9 Product of 5 and 7

10 Sum of 32 and 15

11 Six books at $5 each

12 $5 $\times$ 3

13 Write the smallest number you can using 5, 9, 6, 5. ☐

SET 2 Decimals

Write the decimal that is modelled by the shaded section of each hundredths grid.

1 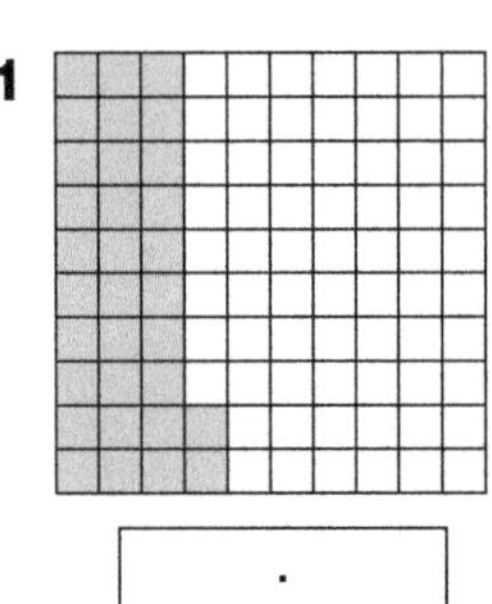

☐ . ☐

2 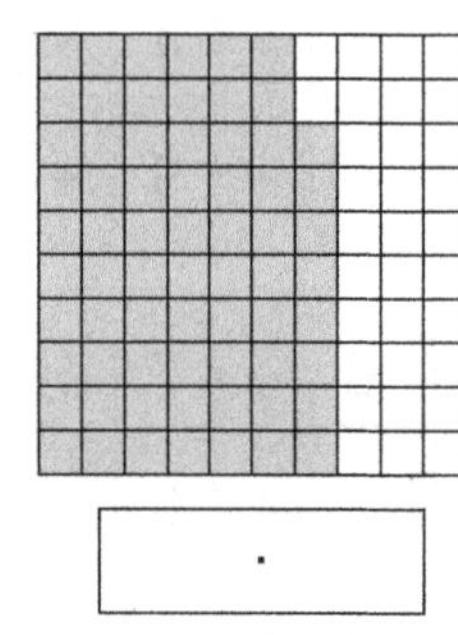

☐ . ☐

3 Write $\frac{15}{100}$ as a decimal.

4 Write $\frac{27}{100}$ as a decimal.

5 Write $\frac{45}{100}$ as a decimal.

6 Write $\frac{17}{100}$ as a decimal.

7 Which is larger, 0.64 or 0.58?

8 Which is larger, 0.22 or 0.75?

9 Which is larger, 0.69 or 0.81?

10 Order these decimals from smallest to largest: 0.38, 0.19, 0.58, 0.72.

Space Combining and splitting shapes

Trace the tangram onto a piece of paper, cut out each shape within the tangram and rearrange the shapes to make the shape below.

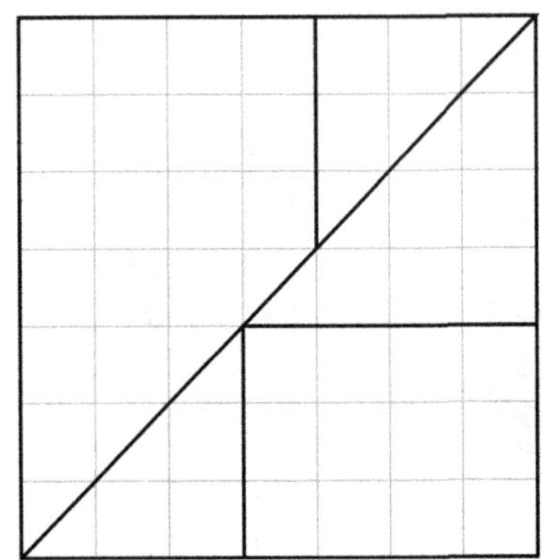

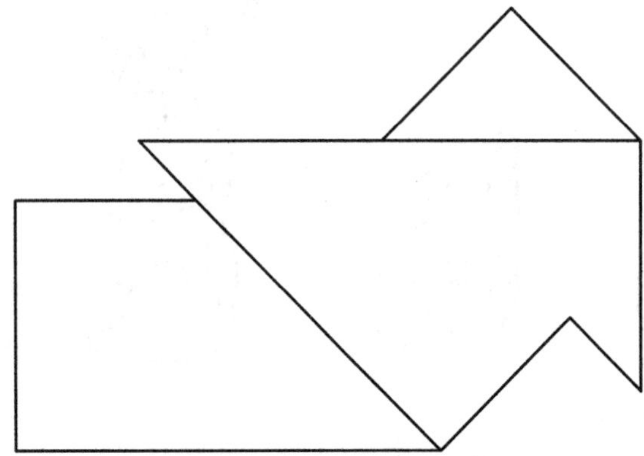

Number and Algebra

SET 3 Multiplication facts

Use any strategy you wish to solve the problems.

1

	× 3
4	
6	
5	
7	
9	
8	

2

	× 5
3	
7	
6	
8	
9	
5	

3

	× 6
1	
2	
4	
9	
6	
5	

Complete the algorithms to identify the winning bingo card.

4 $3 \times 3 =$

5 $10 \times 3 =$

6 $3 \times 4 =$

7 $5 \times 4 =$

8 $6 \times 4 =$

9 $3 \times 6 =$

10 $7 \times 5 =$

11 $7 \times 6 =$

a

9		24		18	
	12		30		35

b

9		20		45	
	12		30		42

c

9		24		18	
	12		30		40

SET 4 Extension

1 Share $1.00 among 5 people.

2 How many 10c coins in $3.00?

3 15 + 4 + 3

4 What is the value of 3 in 2377?

5 $10 – $1.50

6 Write 267 in words.

____

7 How much is $\frac{1}{4}$ of 12 books?

8 What number is 8 more than 1328?

9 How many 20c coins in $4?

10 Which is larger, $\frac{1}{4}$ or $\frac{1}{3}$?

11 297, 300, 303, ____, ____

12 What number is 20 less than 468?

13 Share 27 among 4.

14 How much are 4 books at $6 each?

15 How much are 7 screwdrivers at $4 each?

16 If 6 books cost $24, how much do 5 books cost?

Measurement Metres and centimetres

Draw a line to match each item to an appropriate length.

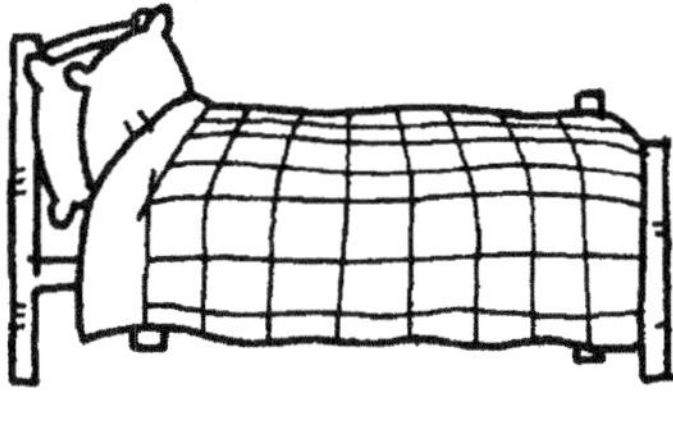

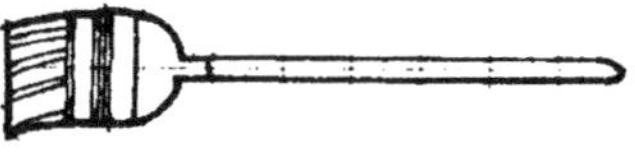

4 m

150 cm

14 cm

2 m

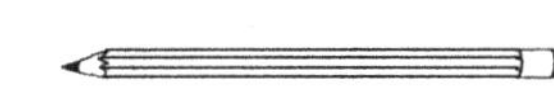

UNIT 35

Number and Algebra

SET 1 Basic

1 6 + 4 + 7

2 7 + 3 + 9

3 18 – 6

4 17 – 8

5 4 × 10

6 5 × 6

7 7 × 3

8 8 × 2

9 Add 27 to 34.

10 Multiply 3 by 5.

11 (6 + 4) × 3

12 How many minutes in 2 hours?

13 Difference between 42 and 20

14 42 – ☐ = 30

15

SET 2 Problems

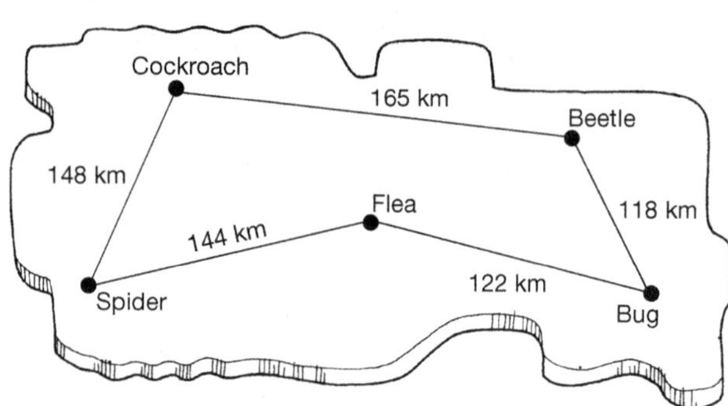

How many kilometres from:

1 Spider to Bug via Flea?

2 Spider to Beetle via Cockroach?

3 Cockroach to Bug via Beetle?

4 Cockroach to Flea via Spider?

5 Cockroach to Flea via Beetle and Bug?

Solve the problems.

6 Ben scored 135 and 146 in two innings. What was his total score for the match? ☐

7 Theo had $95 and Jill had $78. How much money did they have altogether? ☐

8 Son Lee has 5 dozen eggs and Kim has 45 eggs. How many eggs do they have altogether? ☐

Space Hexagons and pentagons

1 Colour all pentagons red.

2 Colour all hexagons green.

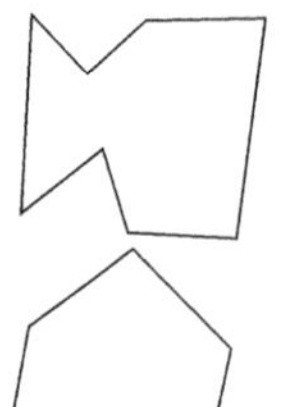

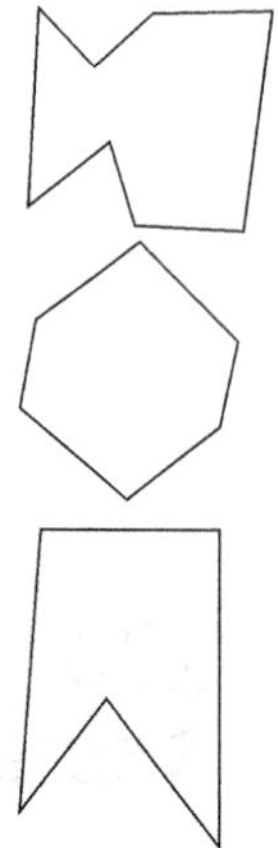

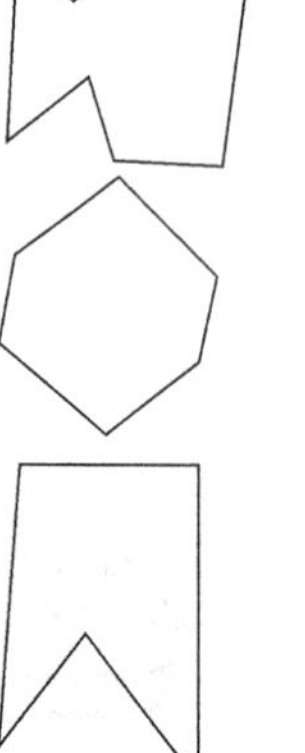

Number and Algebra

SET 3 Missing numbers

Supply the missing numbers or symbols.

1	25	+		=	40
2	33	+		=	50
3	7		2	=	14
4	29		13	=	16
5	88	−		=	50

Working Mathematically

How many ways can you make 20?
One way is shown.

6	50	−	30	=	20
7				=	20
8				=	20
9				=	20
10				=	20
11				=	20

SET 4 Extension

1. Write the largest number you can using 2, 8, 4.
2. How many sides on 7 pentagons?
3. If 6 pens cost \$30, how much do 4 cost?
4. What is half of 48?
5. How many legs on 3 spiders and 2 dogs?
6. 20 less than 496
7. 20 more than 507
8. Value of 4 in 428
9. $(2 \times 10) + 7$
10. $2 \times (10 - 7)$
11. How much are 5 books at \$8 each?
12. 27 + ☐ = 40
13. Write 797 in words.
14. 420 take away 200
15. Subtract 40 from 280.
16. Estimate an answer to 39 + 141.
17. $\frac{1}{3}$ of \$15

Measurement Square centimetres

Record the area of each shape in cm^2.

1 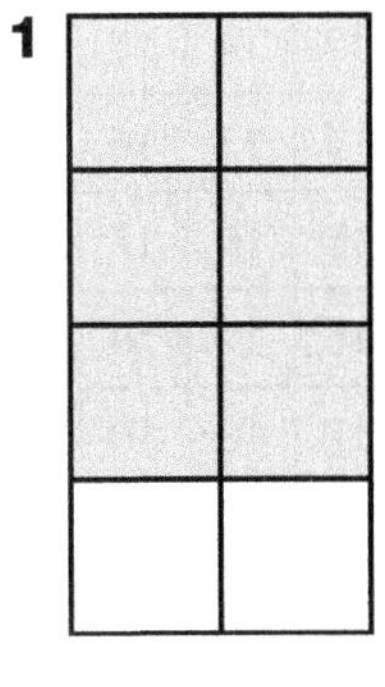

☐ cm^2

2 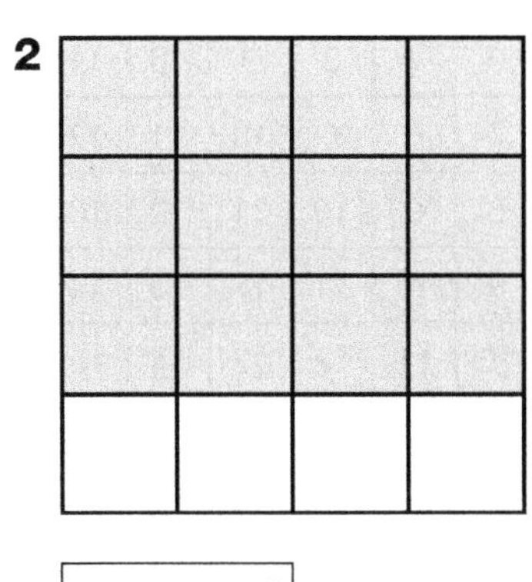

☐ cm^2

3 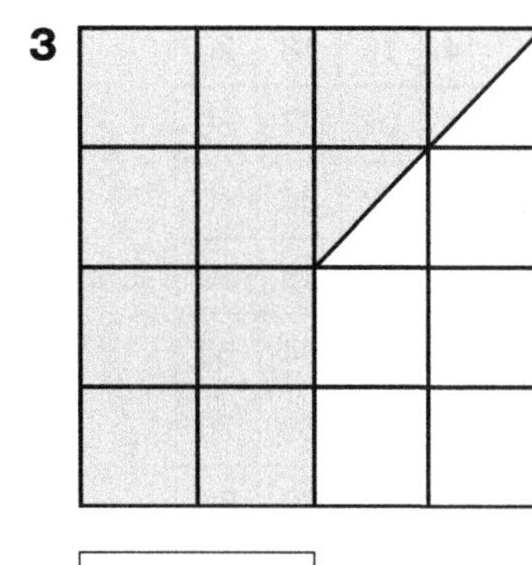

☐ cm^2

4 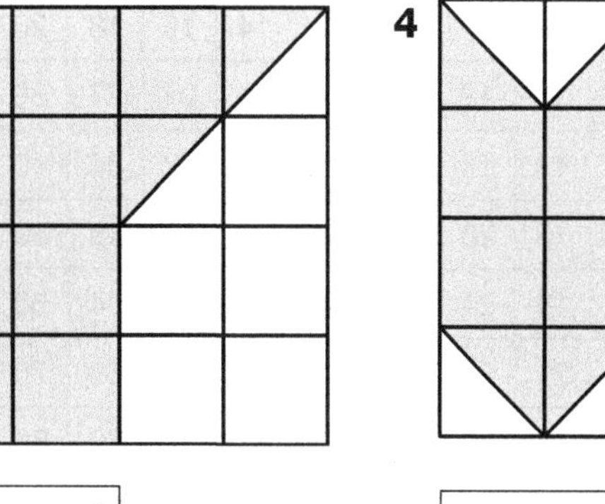

☐ cm^2

Maths helpers

Length

10 millimetres (mm) = 1 centimetre (cm)

100 centimetres (cm) = 1 metre (m)

1000 metres (m) = 1 kilometre (km)

Mass

1000 grams (g) = 1 kilogram (kg)

1000 kilograms (kg) = 1 tonne (t)

Capacity

1000 millilitres (mL) = 1 litre (L)

Time

60 seconds = 1 minute

60 minutes = 1 hour

24 hours = 1 day

7 days = 1 week

14 days = 1 fortnight

12 months = 1 year

52 weeks = 1 year

365 days = 1 year

366 days = 1 leap year

10 years = 1 decade

100 years = 1 century

Months of the year

Thirty days has September, April, June and November. All the rest have thirty-one, except February alone, which has twenty-eight days clear and twenty-nine days each leap year.

Seasons

Summer: December, January, February

Autumn: March, April, May

Winter: June, July, August

Spring: September, October, November

Roman numerals

1 = I

2 = II

3 = III

4 = IV

5 = V

6 = VI

7 = VII

8 = VIII

9 = IX

10 = X

20 = XX

30 = XXX

40 = XL

50 = L

60 = LX

70 = LXX

80 = LXXX

90 = XC

100 = C

500 = D

1000 = M

Multiplication facts

×	0	1	2	3	4	5	6	7	8	9	10
0	0	0	0	0	0	0	0	0	0	0	0
1	0	1	2	3	4	5	6	7	8	9	10
2	0	2	4	6	8	10	12	14	16	18	20
3	0	3	6	9	12	15	18	21	24	27	30
4	0	4	8	12	16	20	24	28	32	36	40
5	0	5	10	15	20	25	30	35	40	45	50
6	0	6	12	18	24	30	36	42	48	54	60
7	0	7	14	21	28	35	42	49	56	63	70
8	0	8	16	24	32	40	48	56	64	72	80
9	0	9	18	27	36	45	54	63	72	81	90
10	0	10	20	30	40	50	60	70	80	90	100

Addition facts

+	2	3	4	5	6	7	8	9	10	11	12
2	4	5	6	7	8	9	10	11	12	13	14
3	5	6	7	8	9	10	11	12	13	14	15
4	6	7	8	9	10	11	12	13	14	15	16
5	7	8	9	10	11	12	13	14	15	16	17
6	8	9	10	11	12	13	14	15	16	17	18
7	9	10	11	12	13	14	15	16	17	18	19
8	10	11	12	13	14	15	16	17	18	19	20
9	11	12	13	14	15	16	17	18	19	20	21
10	12	13	14	15	16	17	18	19	20	21	22
11	13	14	15	16	17	18	19	20	21	22	23
12	14	15	16	17	18	19	20	21	22	23	24

Answers

UNIT 1 Number and Algebra

SET 1

1 8
2 10
3 4
4 9
5 8
6 11
7 19
8 8
9 5
10 6
11 0
12 8
13 8
14 7
15 5

SET 2

1 9
2 17
3 11
4 16
5 18
6 13
7 16
8 13
9 6
10 $20
11 16
12 19
13 9
14 $14
15 36 + 12 = 48

SET 3

1 12, 15, 18
2 20, 22, 24
3 80, 90, 100
4 20, 25, 30
5 22, 26, 30
6 32, 35, 38
7 50, 55, 60
8 70, 75, 80
Rule: Add 5
9 28, 26, 24
Rule: Take away 2
10 41, 46, 51
Rule: Add 5

SET 4

1 5 tens, 5 ones
2 17 + 6
3 $24
4 56
5 Fifteen
6 $38
7 20 + 7 or
14 + 13 or
15 + 12
8 15 + 12 or
14 + 13
9–12 18 + 2, 17 + 3,
16 + 4, 15 + 5

Space

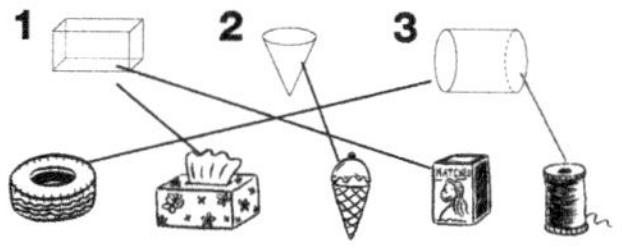

Measurement

Hands on

UNIT 2 Number and Algebra

SET 1

1 9
2 10
3 11
4 11
5 13
6 11
7 10
8 5
9 3
10 3
11 9
12 5
13 12
14 3
15 19

SET 2

1 10
2 12
3 7
4 $9
5 $12
6 3
7 7 km
8 50 cm
9 14
10 4
11 9 min
12 6
13 5
14 11
15 4 pages
16 16

SET 3

1 One hundred and twenty-three
2 Three hundred and sixty-seven
3 237
4 460
5 370, 360
6 5
7 301
8 70
9 10
10
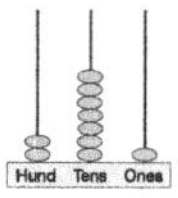

11
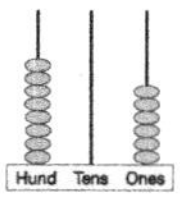

SET 4

1 4
2 42
3 9
4 12
5 Summer
6 31
7 26 January
8 5 June
9 20
10 67
11 3 each
12 19
13 10 km
14 20
15 Hands on

Space

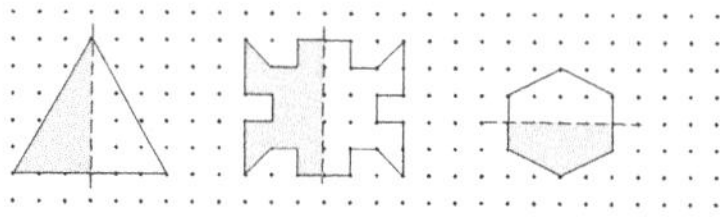

Measurement

Hands on

UNIT 3 Number and Algebra

SET 1

1 11
2 7
3 11
4 9
5 6
6 13
7 4
8 1
9 0
10 6
11 4
12 1
13 6
14 14
15 22

SET 2

1 18
2 38
3 18
4 26
5 36
6 30
7 25
8 35
9 25
10 30
11 20
12 30
13 28
14 7

SET 3

1 2 groups of 2 = 4
2 2 groups of 4 = 8
3 3 groups of 4 = 12
4 4 groups of 5 = 20
5 4 groups of 6 = 24

SET 4

1 6 tens, 9 ones
2 17 + 9
3 51
4 67
5 16
6 1 ten, 9 ones
7 9 tens
8 54
9 41
10 Autumn
11 47
12 3 tens, 6 ones
13 60, 70
14 38
15 $6 \times 4 = 24$ or
$8 \times 3 = 24$
16 $2 \times 12 = 24$ or
$6 \times 4 = 24$

Space

1–9

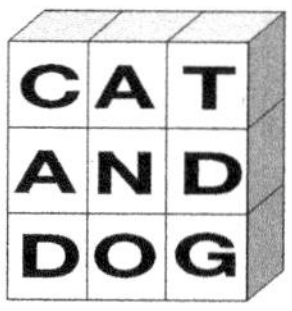

10 CAT AND DOG

Statistics and Probability

1 10
2 13
3 3
4 Holden
5 Toyota and Mazda

Answers

UNIT 4 Number and Algebra

SET 1

1 14
2 13
3 14
4 13
5 14
6 17
7 15
8 14
9 9
10 11
11 12
12 2
13 16
14 8
15 $12

SET 2

1 5 ✓
2 11 ✗
3 7 ✓
4 8 ✓
5 12 ✗
6 12 – 7 = 5
   7 + 5 = 12
7 18 – 5 = 13
   13 + 5 = 18

SET 3

1 2, 4, 6, 8, 10, 12, 14, 16, 18, 20
2 Hands on. Possible answers: 4 × 5, 5 × 4, 10 × 2, 2 × 10

SET 4

1 75
2 5
3 11
4 14
5 132
6 20c
7 5 each
8 27
9 8 tens, 1 one
10 10
11 80
12 176
13 Ninety-two
14 24
15 May
16 Rectangular prism

Space

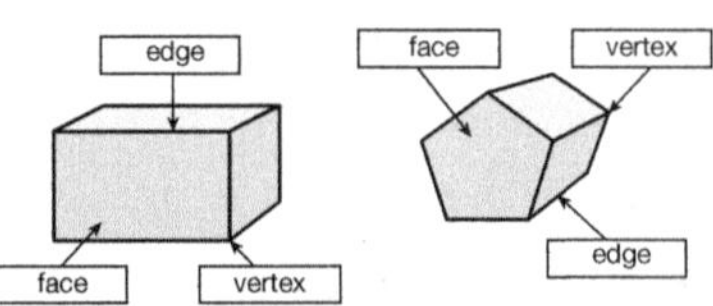

Measurement

1 10 cm
2 12 cm
3 5 cm
4 9 cm
5 13 cm

UNIT 5 Number and Algebra

SET 1

1 11
2 12
3 $12
4 1
5 10
6 12
7 14
8 24
9 7
10 20
11 20
12 40
13 8
14 20
15 20

SET 2

1 • • • •
   • • • odd
2 • • • • •
   • • • • • even
3 • • • • • •
   • • • • • • even
4 • • • • • • • • •
   • • • • • • • • odd
5 odd
6 even
7 odd
8 even
9 even
10 odd
11 even
12 odd
13 even

SET 3

1 13, 7
2 14, 6
3 19, 9
4 17, 5
5 16, 13
6 20, 2
7 18, 9
8 11
9 7
10 $28

SET 4

1 5
2 9 tens, 3 ones
3 4
4 40c
5 Twelve
6 30
7 3
8 September
9 18
10 About 40
11 67
12 8
13 No
14 45
15 6
16 Octagon

Measurement

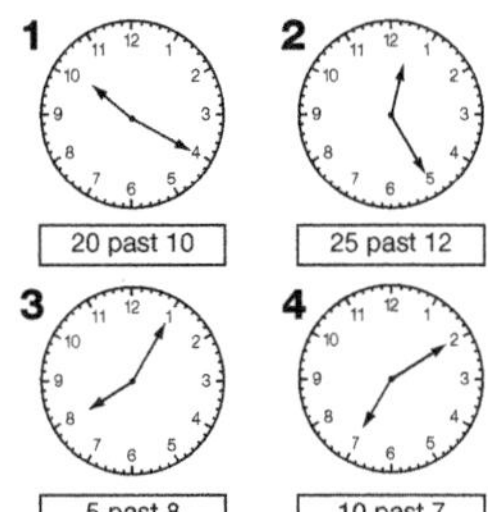

Space

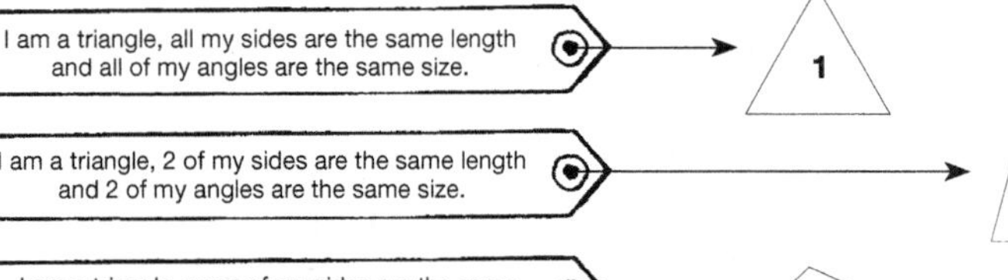

UNIT 6 Number and Algebra

SET 1

1 1
2 4
3 9
4 7
5 9
6 9
7 12
8 4
9 6
10 3
11 14
12 11
13 15c
14 0
15 $20

SET 2

1 500 + 30 + 9
2 600 + 30 + 5
3 200 + 70 + 8
4 300 + 80 + 6
5 700 + 90 + 4
6 400 + 90 + 7
7 <
8 <
9 >
10 <
11 >

SET 3

1 2
2 5
3 4
4 6
5 3
6 5
7 25
8 40
9 50
10 45
11 18 sheep
12 $24
13 $21

SET 4

1 2
2 31
3 6
4 5
5 5c
6 Spring
7 60
8 14
9 20c
10 75
11 43
12 180
13 About 60
14 4 each
15 18, 20, 22, 24, 26, 28
16 24, 26, 28, 30, 32, 34

Space

1 M
2 S
3 V
4 B
5 P

Statistics and Probability

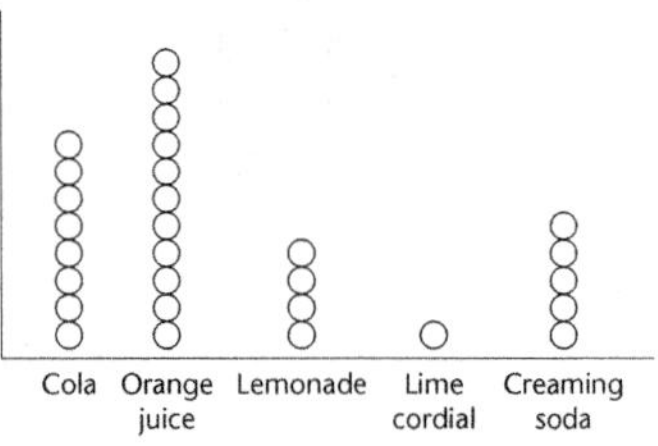

Answers

UNIT 7 Number and Algebra

SET 1

1 9
2 9
3 5
4 10
5 10
6 4
7 3
8 3
9 6
10 14
11 4, 6, 12
12 14
13 14
14 9
15 9

SET 2

1 3
2 4
3 6
4 2
5 8
6 4
7 16
8 2
9 1

SET 3

1 19
2 25
3 31
4 24
5 38
6 24
7 46
8 93
9 6
10 $20
11 31 cm
12 85
13 35
14 40

SET 4

1 $5 each
2 81
3 64
4 16
5 Sixteen
6 No
7 69
8 24
9 9
10 May
11

Space

Hands on

Measurement

1 Bucket
2 Glass or 1 L milk carton

UNIT 8 Number and Algebra

SET 1

1 6
2 8
3 11
4 9
5 1
6 4
7 6
8 7
9 0
10 28
11 32
12 20
13 50
14 11
15 50c

SET 2

1 5
2 50
3 4
4 40
5 7
6 70
7 4
8 40
9 400
10 200
11 30 kg
12 $500

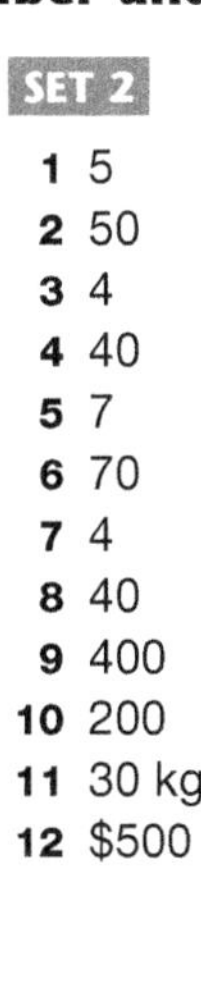

SET 3

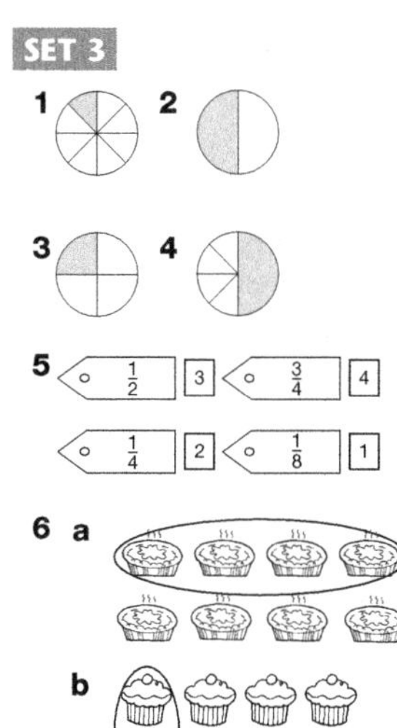

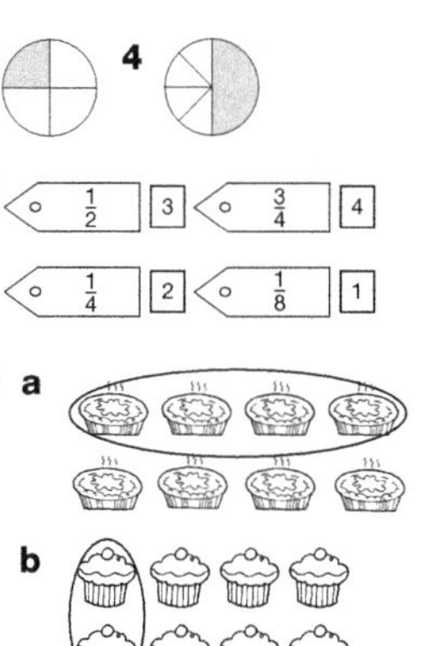

SET 4

1 12
2 15
3 6 each
4 March
5 190
6 Seventeen
7 14
8 86
9 41
10 31
11 30
12 2
13 About 60
14 Hands on

Space

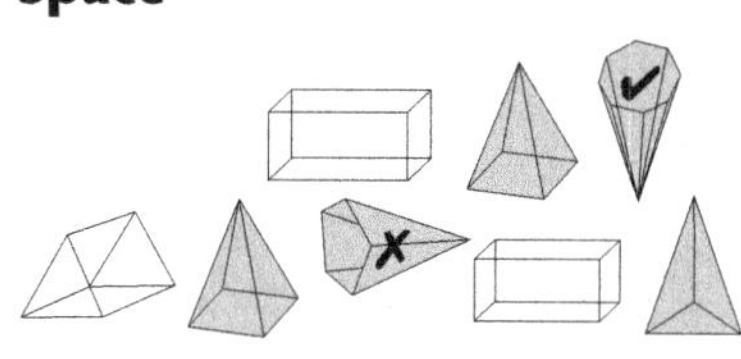

Measurement

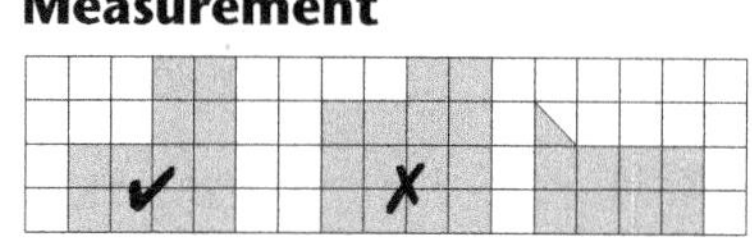

UNIT 9 Number and Algebra

SET 1

1 3
2 6
3 9
4 9
5 9
6 9
7 12
8 18
9 5c
10 1
11 15
12 10
13 12
14 24
15 15

SET 2

1 27 + 32 becomes 50 + 9 = 59
2 44 + 23 becomes 60 + 7 = 67
3 35 + 17 becomes 40 + 12 = 52
4 58 + 22 becomes 70 + 10 = 80
5 45
6 33
7 59
8 48
9 51
10 68

SET 3

1 4
2 2
3 12
4 4
5 20
6 40
7 8
8 16
9 20
10 40
11 Any multiplication or division fact with a product less than 40

SET 4

1 2
2 28
3 80
4 30
5 Twenty-three
6 4 each
7 February
8 24
9 95
10 5 each
11 8 tens, 5 ones
12 4
13 August
14 20
15 Hands on. One example

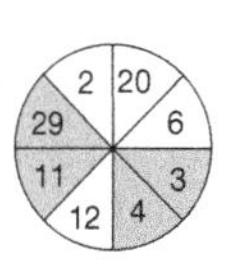

Statistics and Probability

1 5
2 9
3 2
4 29
5 Blond and black
6 Hands on

Measurement

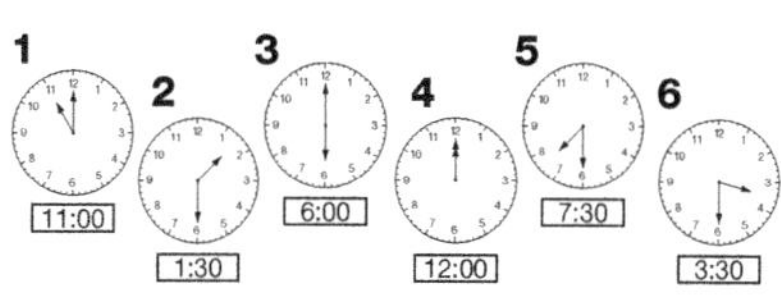

Answers

UNIT 10 Number and Algebra

SET 1

1 13
2 $2
3 12
4 12
5 10
6 24
7 16
8 5
9 13
10 16
11 95c
12 6
13 30
14 20
15 $14

SET 2

1 17 – 6 = 11
6 + 11 = 17
2 23 – 9 = 14
9 + 14 = 23
3 35 – 31 = 4
31 + 4 = 35
4 57 – 52 = 5
52 + 5 = 57
5 27 – 18 = 9
18 + 9 = 27
6 45 – 32 = 13
32 + 13 = 45
7 40 + 20 + 10 = 70
30 + 30 + 10 = 70
30 + 20 + 20 = 70

SET 3

1 $\frac{2}{8} < \frac{5}{8}$
2 $\frac{3}{8} > \frac{2}{8}$
3 $\frac{6}{8} > \frac{5}{8}$
4 $\frac{1}{2} = \frac{4}{8}$
5 $\frac{4}{8} < \frac{6}{8}$
6 $\frac{7}{8} > \frac{3}{8}$
7 $\frac{1}{2} < \frac{2}{2}$
8 $\frac{4}{4} = \frac{2}{2}$
9 $\frac{8}{8} > \frac{3}{4}$
10 $\frac{1}{2} < \frac{3}{4}$

Possible solutions

11 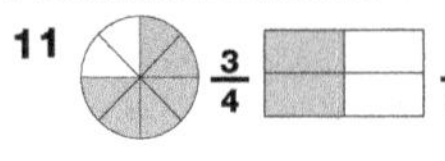

SET 4

1 8 tens, 2 ones
2 3 + 8
3 30
4 90c
5 18
6 2 February
7 35
8 10
9 6
10 Ninety-five
11 62
12 2 hund, 3 tens, 4 ones
13 18
14 Twenty-eight
15 93

Space

1 Blue
2 Red
3 Blue
4 Red
5 Red

Measurement

Possible answers:

	Item	Taller than 1 m	About the same as 1 m	Shorter than 1 m
1	Broom			
2	Chair			
3	Refrigerator			
4	Rubbish bin			
5	Shovel			
6	Telephone booth			

UNIT 11 Number and Algebra

SET 1

1 10
2 10
3 20
4 20
5 3
6 4
7 5
8 6
9 12
10 24
11 16
12 20
13 14
14 15
15 $12

SET 2

1 34, 44, 54, 64, 74, 84
2 150, 160, 170, 180, 190, 200
3 77, 87, 97, 107, 117, 127
4 73, 63, 53, 43, 33, 23
5 80, 70, 60, 50, 40, 30
6 235, 225, 215, 205, 195, 185
7 64, 74
8 152, 162
9 84, 94
10 155, 165
11 180, 190

SET 3

1
2 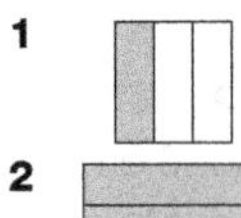
3 
4 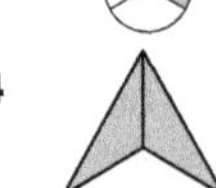
5
6 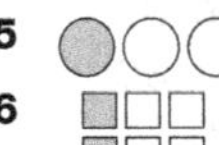
7 No
8 No
9 No

SET 4

1 16
2 11, 12, 17, 19
3 58
4 One hundred and eighty-seven
5 325
6 175
7 15
8 $1.50
9 21
10 200
11 46
12 50c
13 13
14 3
15 5
16 Pentagon

Number and Algebra

1 12, 14, 16, 18
Rule: Add 2
2 15, 18, 21, 24
Rule: Add 3
3 35, 40, 45, 50
Rule: Add 5
4 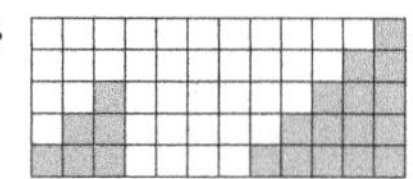
The next number of blocks is 21.

Measurement

Hands on

UNIT 12 Number and Algebra

SET 1

1 20
2 17
3 18
4 16
5 13
6 11
7 17
8 18
9 18
10 36
11 10
12 14
13 4
14 10
15 $15

SET 2

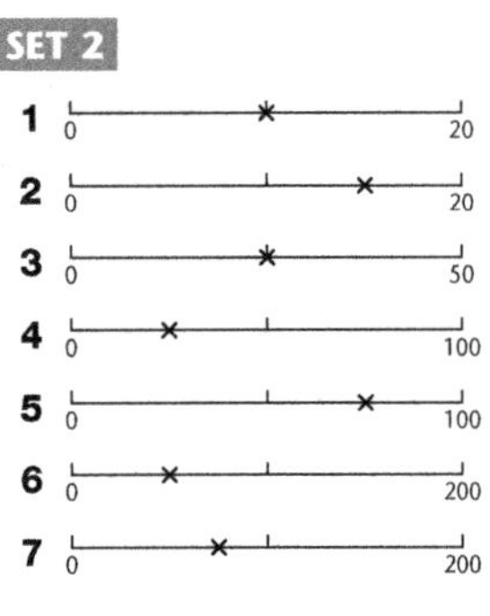

SET 3

1 30
2 50 plants
3 40
4 70
5 0
6 40
7 80
8 9, 15, 18, 6, 27, 24, 21, 12
9 6, 10, 12, 4, 18, 16, 14, 8
10 30, 50, 60, 20, 90, 80, 70, 40
11 15, 25, 30, 10, 45, 40, 35, 20

SET 4

1 Summer
2 1, 2, 3, 5, 7
3 13
4 9
5 58
6 Sixty-four
7 89
8 12 each
9 15
10 $24
11 28
12 100
13 120
14 25
15 114

Space

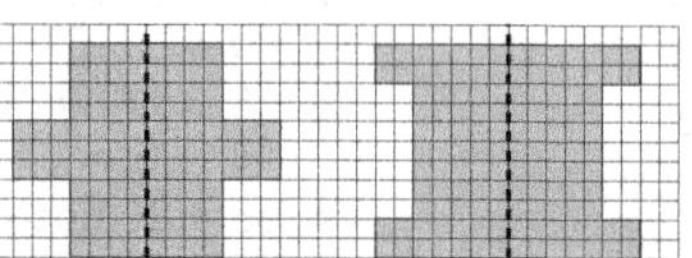

Statistics and Probability

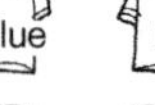

Answers

UNIT 13 Number and Algebra

SET 1

1 9
2 11
3 16
4 $4
5 19
6 18
7 20
8 21
9 13
10 5
11 12
12 14
13 20
14 6
15 8

SET 2

1 23, 32, 64, 43, 55
2 12, 21, 53, 32, 44
3 35, 54, 21, 55, 43
4 16, 35, 2, 36, 24
5 51
6 30

SET 3

1 Any 3 parts
2 Any 7 parts
3 Any 4 parts
4 Any 3 parts
5 Any 2 circles
6 Any 9 squares
7 True
8 False
9 True
10 True

SET 4

1 82
2 73
3 3
4 15
5 $19
6 37
7 22
8 24
9 3 each
10 20
11 August
12 65c
13 60
14 20
15 September
16

Space

1 Hands on
2 Hands on
3

Angle	Smaller than a right angle	Right angle	Larger than a right angle

Measurement

1 8 L
2 3 L
3 2 L
4 1 L
5 5
6 2

UNIT 14 Number and Algebra

SET 1

1 10
2 11
3 14
4 9
5 5
6 12
7 22
8 17
9 15
10 19
11 14
12 Yes
13 7
14 20
15 7th

SET 2

1 9 + 8 becomes 9 + 1 + 7 = 17
2 18 + 7 becomes 18 + 2 + 5 = 25
3 43 + 9 becomes 43 + 7 + 2 = 52
4 64 + 8 becomes 64 + 6 + 2 = 72
5 36 + 5 becomes 36 + 4 + 1 = 41
6 45 + 7 becomes 45 + 5 + 2 = 52
7 44 km
8 45 km
9 67 km
10 42 km
11 47 km

SET 3

1 6
2 4
3 12
4 2
5 3
6 8
7 5
8 3
9 5
10 5
11 5

SET 4

1 213
2 30
3 406
4 25
5 25
6 24
7 2 hund, 7 tens, 6 ones
8–11 Hands on
12 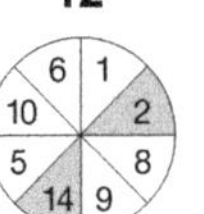

13 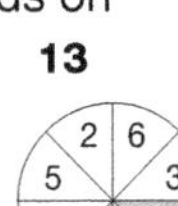

Space

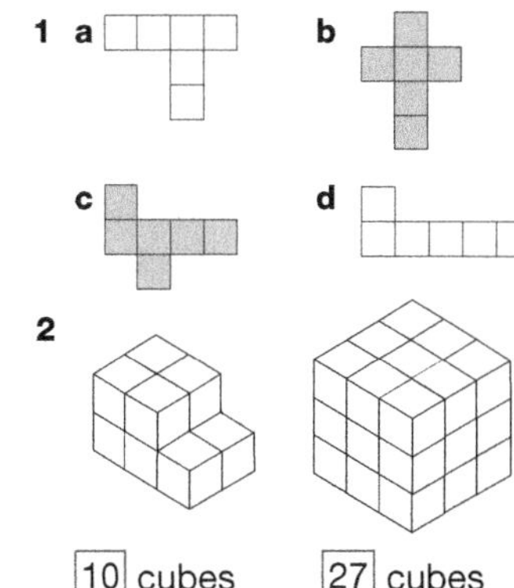

10 cubes 27 cubes

Statistics and Probability

1 John, Tim, Kelly
2 Jessica, Con, Abdul, Sam
3 Kim, Greg, Maria, Soula

UNIT 15 Number and Algebra

SET 1

1 13
2 15
3 14
4 23
5 10
6 9
7 8
8 12
9 20
10 14
11 12
12 12
13 325
14 13
15 12

SET 2

1 60
2 80
3 90
4 80
5 90
6 90
7 60 km
8 80 km
9 80 km

SET 3

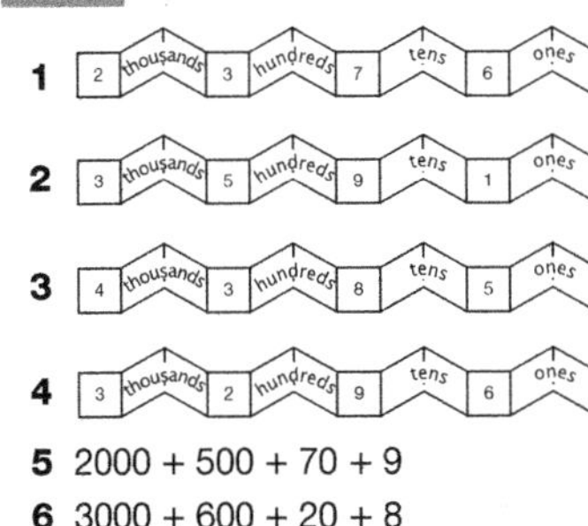

5 2000 + 500 + 70 + 9
6 3000 + 600 + 20 + 8
7 4000 + 100 + 90 + 6
8 7000 + 800 + 90 + 6
9 9000 + 400 + 90 + 4
10 1341, 1437, 3471

SET 4

1 45
2 24
3 21
4 About 110
5 360
6 15
7 25
8 327
9 8 each
10 300
11 One hundred and seventy-two
12 40
13 2
14 $1 or 100c
15 14
16 15

Number and Algebra

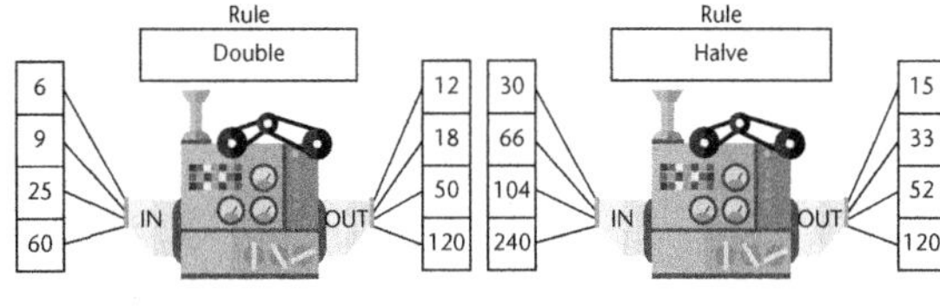

Measurement

1 
2 
3 
4 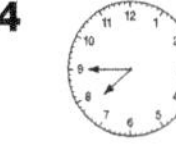
5 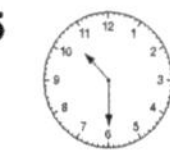

Answers

UNIT 16 Number and Algebra

SET 1

1 23
2 16
3 50
4 20
5 22
6 20
7 20
8 14
9 20
10 15
11 15
12 20
13 4
14 2
15 $20

SET 2

1

9	4	5
2	6	10
7	8	3

2

7	2	9
8	6	4
3	10	5

3

9	4	11
10	8	6
5	12	7

4

14	9	10
7	11	15
12	13	8

SET 3

1 6
2 10
3 14
4 18
5 12
6 30
7 50
8 70
9 90
10 60
11 15
12 25
13 35
14 45
15 30
16 40
17 Hands on
$1 \times 24, 24 \times 1$
$2 \times 12, 12 \times 2$
$3 \times 8, 8 \times 3$
$4 \times 6, 6 \times 4$

SET 4

1 1300
2 56
3 9 tens, 5 ones
4 1 hund, 6 tens, 2 ones
5 627
6 10 each
7 135
8 Thirty-nine
9 15
10 200, 300
11 $2.50
12 239
13 7th
14 3
15 $1.20

Statistics and Probability

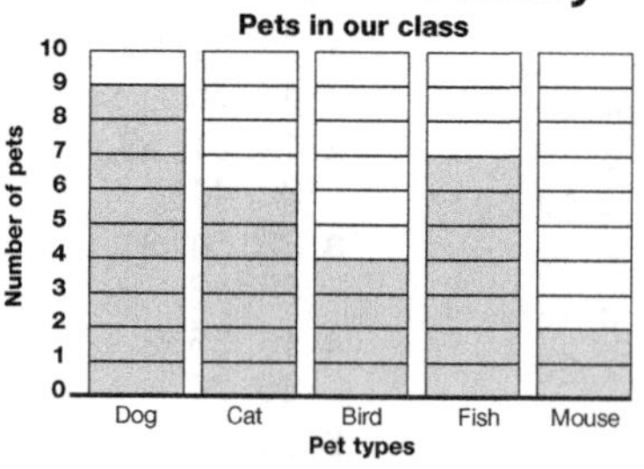

Measurement

Scales **a** and **d** are correct.

UNIT 17 Number and Algebra

SET 1

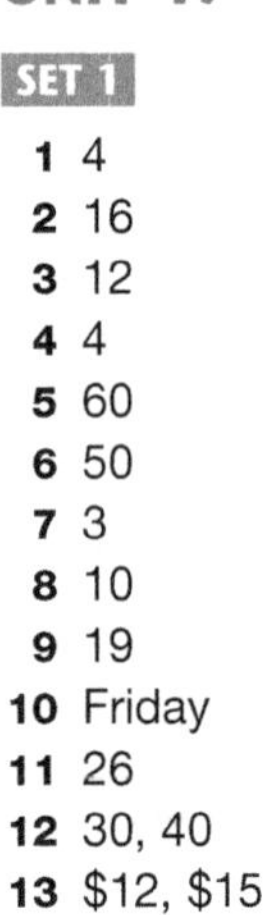

1 4
2 16
3 12
4 4
5 60
6 50
7 3
8 10
9 19
10 Friday
11 26
12 30, 40
13 $12, $15

SET 2

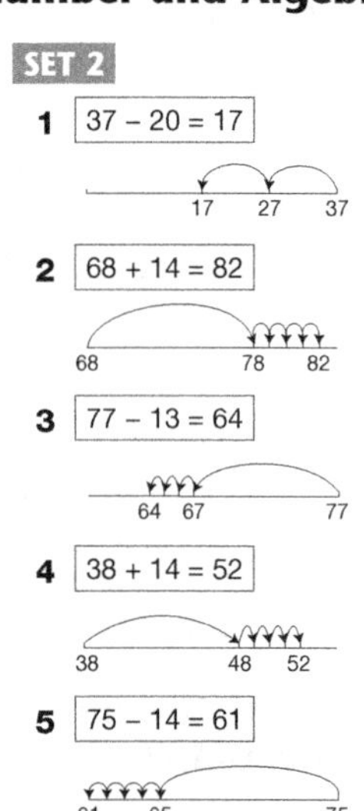

1 37 – 20 = 17

2 68 + 14 = 82

3 77 – 13 = 64

4 38 + 14 = 52

5 75 – 14 = 61

SET 3

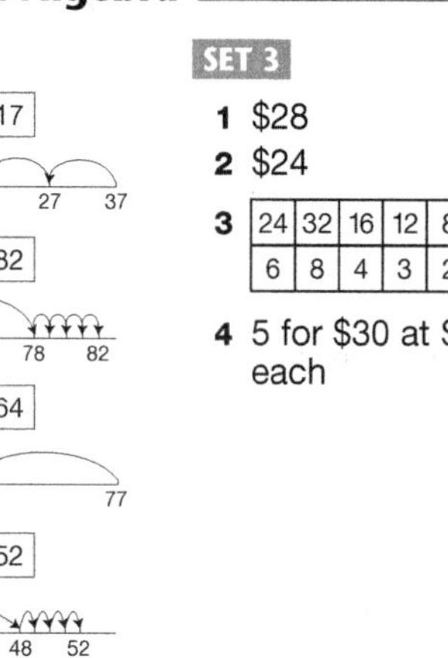

1 $28
2 $24
3

24	32	16	12	8	40	4
6	8	4	3	2	10	1

4 5 for $30 at $6 each

SET 4

1 17
2 1000
3 2 dozen
4 754
5 2300
6 30
7 0
8 About 280
9 89
10 12
11 5 each
12 155, 161
13 $1.50
14 150 cm
15 7 June

Space

1 to 5

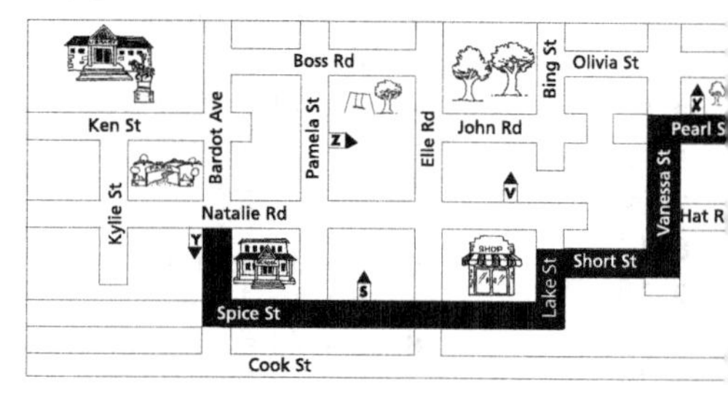

6 Letter Y

Measurement

1 20 cm
2 Hands on

UNIT 18 Number and Algebra

SET 1

1 8
2 18
3 28
4 38
5 7
6 11
7 10
8 8
9 6
10 28
11 45
12 32
13 12
14 35
15 $28

SET 2

1 51
2 91
3 80
4 82
5 82
6 82
7 73
8 78
9 68
10 77
11 57
12 77
13 79
14 69
15 $89
16 $53

SET 3

1 $60, $55, $50
(Rule: Subtract $5)
2 $28, $32, $36
(Rule: Add $4)
3 28, 35, 42, 49
4 46, 42, 38, 34
5 36, 45, 54, 63
6 89, 97, 105, 113
7 60, 70, 80, 90
Rule: Add 10
8 21, 24, 27, 30
Rule: Add 3

SET 4

1 8
2 3
3 3
4 24
5 $35
6 473
7 971
8 179
9 976
10 $1
11 16 + 8 + 4 + 2 + 1 = 31

Space

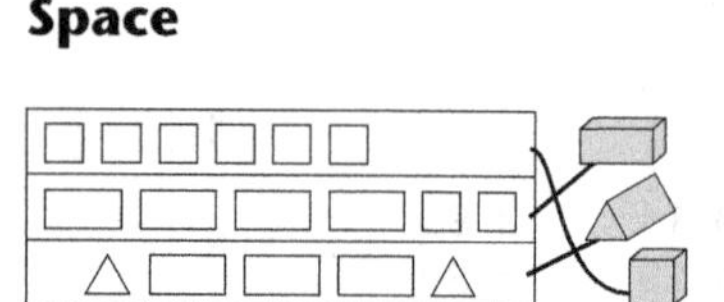

Measurement

Hands on

UNIT 19 Number and Algebra

SET 1

1 13
2 $4
3 8
4 $16
5 5
6 20
7 12
8 11
9 3
10 4
11 24
12 19
13 35
14 90
15 7

SET 2

1 600, 700, 800
2 450, 550, 650
3 600, 500, 400
4 533, 633, 733
5 580, 480, 380
6 471, 571, 671
7 179, 330, 210, 903, 700
8 200, 412, 20, 276, 888
9 105, 205, 305, 405, 505

SET 3

1		2		3	
1	20	2	9	3	2
	30		21		4
	25		18		8
	35		24		18
	45		27		12
	40		15		10

4 9
5 18
6 12
7 15
8 21
9 24
10 30
11 45

The winning card is **b**.

SET 4

1 $16
2 35
3 51
4 240
5 12 each
6 $95
7 675
8 349
9 7
10 241, 247
11 4 5 6   4 6 5
5 4 6   5 6 4
6 4 5   6 5 4

Space

1
2

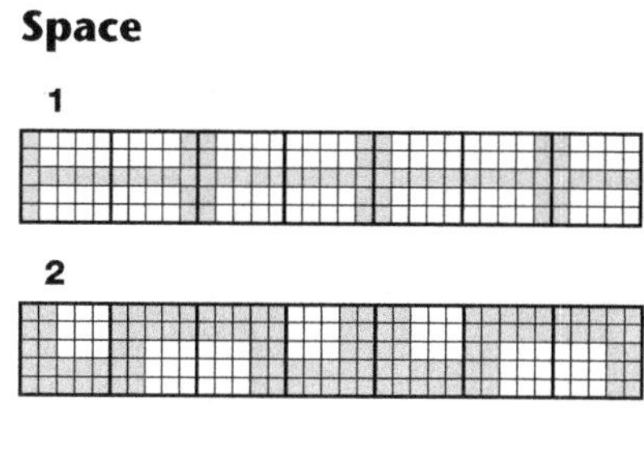

Statistics and Probability

1 'Likely' is the best choice in this instance.
2 Hands on

UNIT 20 Number and Algebra

SET 1

1 27
2 29
3 32
4 35
5 54
6 14
7 10
8 14
9 15
10 12
11 27
12 12
13 90
14 75
15 $14

SET 2

1 19
2 29
3 15
4 18
5 17
6 48
7 27
8 17
9 58
10 37
11 25
12 34

SET 3

1 24
2 8
3 40
4 8
5 10
6 40
7 21
8 7
9 Secret words: ICE CREAM
10 3
11 4 with 4 left over

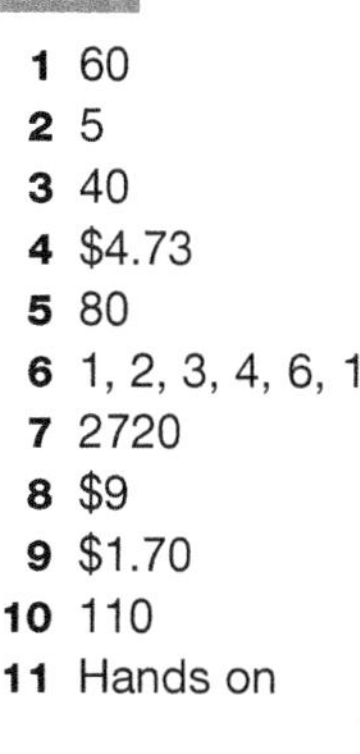

SET 4

1 60
2 5
3 40
4 $4.73
5 80
6 1, 2, 3, 4, 6, 12
7 2720
8 $9
9 $1.70
10 110
11 Hands on

Space

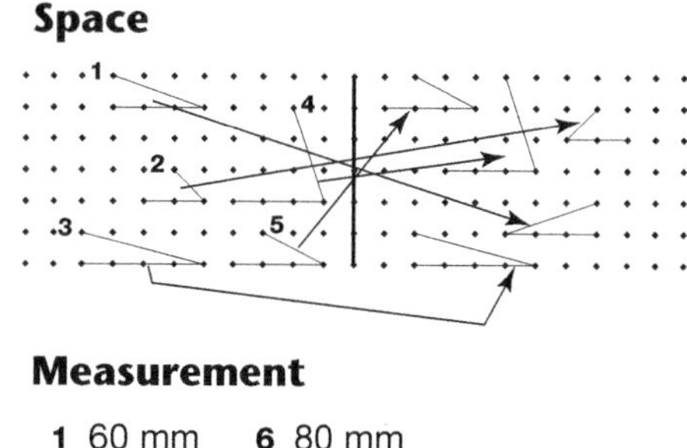

Measurement

1 60 mm
2 90 mm
3 40 mm
4 50 mm
5 20 mm
6 80 mm
7 100 mm
8 70 mm
9 10 mm
10 5 mm

UNIT 21 Number and Algebra

SET 1

1 3
2 18
3 40
4 20
5 42
6 15
7 42
8 32
9 10c
10 2
11 59
12 40
13 13
14 31
15 11

SET 2

1 135
2 59
3 142
4 126
5 $139
6 161
7 116
8 398
9 989
10 875
11 523
12 712
13 573

SET 3

1 $6.35
2 $8.55
3 $7.15
4 $5.85
5 $6.65
6 $7.55

SET 4

1 2
2 0c
3 6
4 87, 90
5 $4
6 $20
7 Yes
8 23
9 7000
10 2000
11 300
12 60c
13 12 lollies
14 232, 234, 236, 238

Number and Algebra

1 80
2 100
3 60
4 62
5 82
6 182
7 257
8 393
9 423

Measurement

Hands on

Answers

UNIT 22 Number and Algebra

SET 1

1 9
2 24
3 31
4 41
5 24
6 10
7 19
8 18
9 24
10 18
11 30
12 60
13 24
14 29
15 21

SET 2

1 819
2 258
3 428
4 636
5 717
6 348
7 257
8 525
9 346
10 357
11 708
12 255
13 $322

SET 3

1 3
2 5
3 6
4 2
5 1
6 1
7 $8
8 $4
9 $2

SET 4

1 February
2 61
3 43, 46
4 93
5 10 o'clock
6 250
7 24
8 $5.30
9 $1.85
10 8 each
11 72
12 48
13 11 pm
14 $\frac{1}{2}$
15 2000
16 26

Space

1 $\frac{1}{4}$
2 $\frac{1}{2}$
3 $\frac{3}{4}$
4 $\frac{1}{2}$

Measurement

1 7:20
2 6:25
3 9:10
4 9:50
5 5:00
6 11:05

UNIT 23 Number and Algebra

SET 1

1 12
2 100
3 $6
4 5c
5 60c
6 5c
7 16
8 0
9 $17
10 10c
11 10c
12 27
13 $1.25
14 45
15 8

SET 2

1 × 5: 2 → 10, 1 → 5, 3 → 15, 6 → 30, 4 → 20, 8 → 40, 7 → 35, 5 → 25
2 × 3: 2 → 6, 1 → 3, 3 → 9, 6 → 18, 4 → 12, 8 → 24, 7 → 21, 5 → 15
3 × 4: 2 → 8, 1 → 4, 3 → 12, 6 → 24, 4 → 16, 8 → 32, 7 → 28, 5 → 20

4 12
5 6
6 8
7 90
8 0
9 48
10 30
11 3
12 50
13 24
14 5
15 24

SET 3

1	36 + 46	becomes	40	+	42	=	82
2	46 + 59	becomes	50	+	55	=	105
3	29 + 38	becomes	30	+	37	=	67
4	52 + 37	becomes	50	+	39	=	89
5	64 + 88	becomes	70	+	82	=	152
6	168 + 27	becomes	170	+	25	=	195
7	66 − 38	becomes	68	−	40	=	28
8	72 − 46	becomes	76	−	50	=	26
9	91 − 59	becomes	92	−	60	=	32
10	85 − 37	becomes	88	−	40	=	48
11	151 − 46	becomes	155	−	50	=	105
12	182 − 69	becomes	183	−	70	=	113

SET 4

1 $1.70
2 $1.85
3 $1.51
4 $4.33
5 50c each
6 300
7 $2.70
8 90c
9 Apple pie
$5 – $1.30 = $3.70

Space

1 Square
2 Star
3 Pentagon
4–5

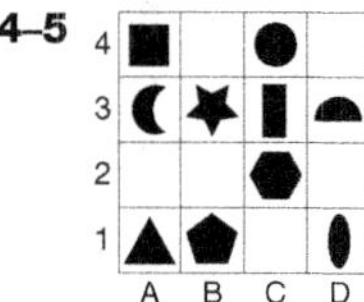

6 C4

Statistics and Probability

1 Apple
2 Pear
3 Orange and banana
4 Apple

UNIT 24 Number and Algebra

SET 1

1 14
2 11
3 23
4 13
5 13
6 13
7 25
8 12
9 50
10 30
11 35
12 20
13 27
14 8
15 30

SET 2

1 $23 + 40 - 2 = 61$
2 $32 + 30 - 1 = 61$
3 $18 + 50 - 1 = 67$
4 $25 + 40 - 1 = 64$
5 $27 + 50 - 2 = 75$
6 $46 + 20 - 1 = 65$
7 $48 + 50 - 1 = 97$
8 $53 + 30 - 2 = 81$
9 58
10 79
11 95
12 83
13 108

SET 3

1 3
2 3
3 6
4 6
5 5

SET 4

1 $45
2 870
3 About 170
4 $3
5 270
6 Two hundred and seventy-two
7 $2.40
8 300
9 30
10 5
11 Hands on. Possible solutions:
3 T-shirts
10 ties
6 pairs of socks
2 T-shirts, 2 pairs of socks
5 ties, 1 pair of socks, 1 T-shirt
1 T-shirt, 4 pairs of socks
5 ties, 3 pairs of socks

Statistics and Probability

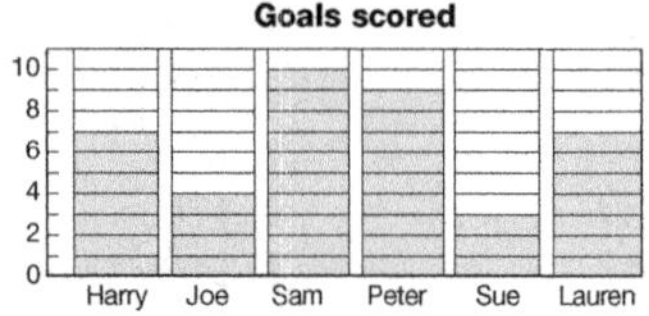

Measurement

75 kg

UNIT 25 Number and Algebra

SET 1

1 25
2 0
3 39
4 15
5 1000
6 28
7 6
8 10
9 20
10 90
11 36
12 40
13 2
14 10
15 $18

SET 2

1 26 km
2 43 km
3 33 km
4 32 km
5 45 km
6 81
7 $88
8 51

SET 3

Draw a line to match the fractions to the number line.

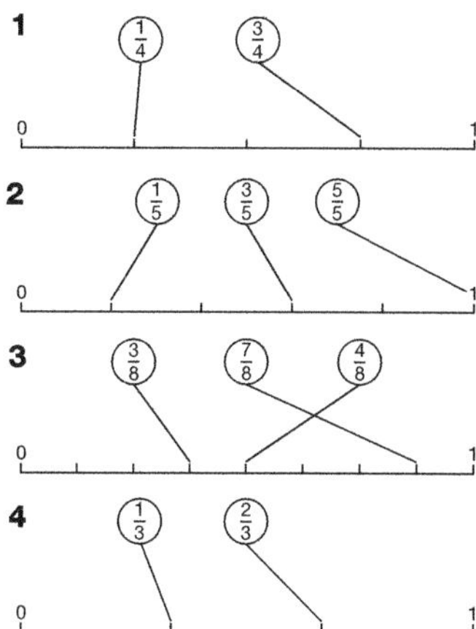

SET 4

1 36
2 250
3 15
4 5
5 18
6 48
7 4 each
8 42
9 189
10 $20 each
11 65
12 48
13 366
14 $17.00
15 2 km
16 145

Shape

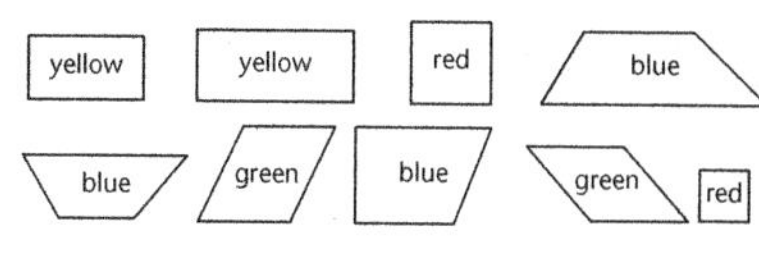

Measurement

Hands on

UNIT 26 Number and Algebra

SET 1

1 30
2 0
3 27
4 7
5 1000
6 18
7 8
8 6
9 17
10 89
11 24
12 31
13 2
14 10
15 61
16 3 × 7

SET 2

1 443
2 623
3 320
4 322
5 242
6 214
7 802
8 915
9 712
10 230
11 about 400
12 about 100
13 about 400
14 about 300

SET 3

1 3674

3 thousands 6 hundreds 7 tens 4 ones

2 3842

3 thousands 8 hundreds 4 tens 2 ones

3 2357

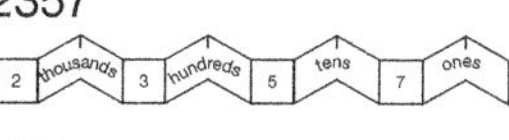

4 700
5 70
6 500
7 3000
8 9
9 3000

SET 4

1 45
2 250
3 4
4 5
5 18
6 20c
7 12 each
8 180
9 $\frac{1}{4}$
10 91
11 633
12 18
13 365
14 $64
15 2400

Shape

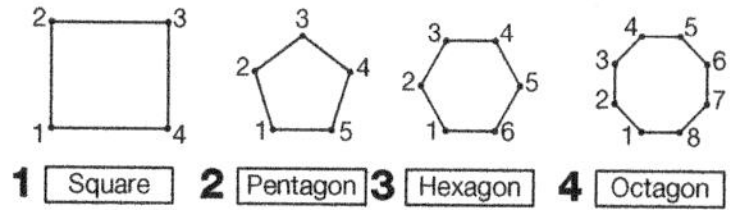

Measurement

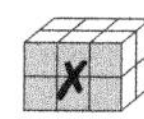
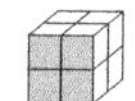
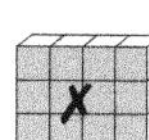

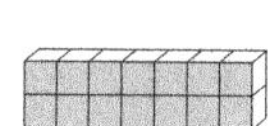

UNIT 27 Number and Algebra

SET 1

1 47
2 3
3 19
4 10c
5 40
6 Yes
7 20
8 32
9 149
10 39
11 12
12 Yes
13 11
14 24
15 18

SET 2

1 0, 30, 6, 9, 21, 18, 24
2 0, 40, 8, 12, 28, 24, 32
3 0, 100, 20, 30, 70, 60, 80
4 0, 50, 10, 15, 35, 30, 40
5 $35
6 $42
7 $24
8 20, 25, 30, 35, 40
9 16, 20, 24, 28, 32
10 12, 15, 18, 21, 24
11 36, 45, 54, 63, 72
12 24, 30, 36, 42, 48
13 True

SET 3

1 2
2 5
3 10
4 4
5 5
6 2
7 4
8 3
9 5
10 4
11 5

SET 4

1 $5.80
2 202 km
3 20 hours
4 171 cm
5 $7.50
6 $119
7 Possible solution

1
2 5 6
7

Shape

1 Start at ■. Travel east 4 spaces.
2 Travel north 3 spaces.
3 Travel west 2 spaces.
4 Travel north 2 spaces.
5 Travel east 7 spaces.
6 Travel south 5 spaces.

Statistics and Probability

Hands on. Possible solutions:

- 5 red marbles, 4 green and 0 blue
- 6 red marbles, 3 green and 0 blue
- 7 red marbles, 2 green and 0 blue
- 8 red marbles, 1 green and 0 blue

Answers

UNIT 28 Number and Algebra

SET 1

1 13
2 6
3 7
4 62
5 43
6 90
7 12
8 24
9 3
10 30
11 42
12 48
13 2
14 4
15 6

SET 2

1 ✔
2 ✔
3 ✘ You would receive $34.
4 ✔
5 ✘ They cost $129.

SET 3

1 15, 15
2 15, 15
3 25, 25
4 21, 21
5 35, 35
6 23, 23
7 38, 38
8 69, 69
9 No
10 10, 10
11 12, 12
12 15, 15
13 20, 20
14 35, 35
15 24, 24
16 No

SET 4

1 4
2 30c
3 73, 77
4 $12
5 No
6 24
7 177
8 83, 85, 87, 89
9 750
10 471
11 4
12 $2.86
13 1, 2, 4, 8, 16
14 16
15 $5.40
16 6 r 3

Space

1 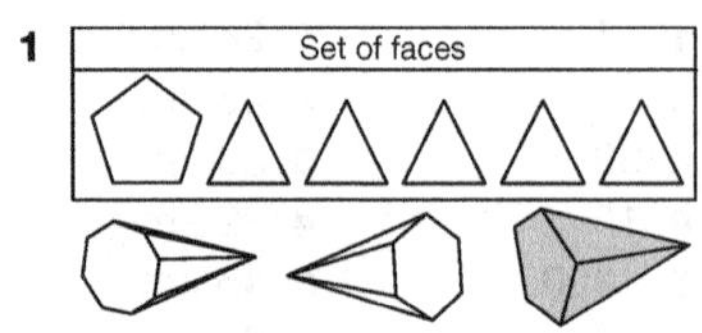

Measurement

UNIT 29 Number and Algebra

SET 1

1 7
2 5c
3 9
4 24
5 12
6 11
7 36
8 70
9 50
10 29
11 15c
12 28
13 500
14 7
15 60

SET 2

1 740
2 792
3 785
4 422
5 854
6 631

	Item	Value
7	Watch	$290
8	Ring	$3499
9	Necklace	$4999
10	Bracelet	$5499
	Total	$14287

SET 3

1 2
2 4
3 8
4 6 apples
5 3 apples

SET 4

1 786
2 991
3 106
4 317
5 4 r 3
6 247
7 400, 425
8 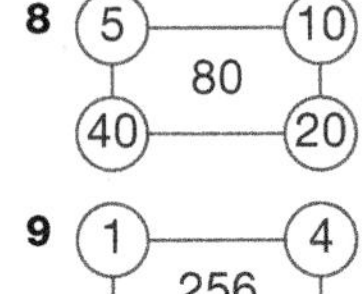

9 1, 4, 256, 64, 16
10 Hands on

Space

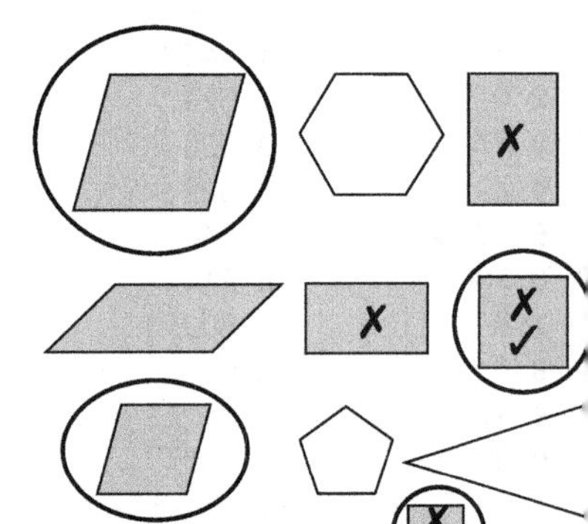

Statistics and Probability

1 Yellow
2 Yes
3 **a** Blue **b** Yes

UNIT 30 Number and Algebra

SET 1

1 6
2 $13
3 20
4 30
5 18
6 33
7 16
8 36
9 21
10 45
11 32
12 24
13 18
14 36
15 5

SET 2

1 7398
2 5992
3 7875
4 9523
5 5782
6 6491
7 $1842
8 $8536

SET 3

1 $\frac{1}{2}$ and $\frac{1}{2}$
2 $\frac{3}{4}$ and $\frac{1}{4}$
3 $\frac{1}{3}$ and $\frac{2}{3}$
4 $\frac{3}{8}$ and $\frac{5}{8}$
5 $\frac{4}{7}$ and $\frac{3}{7}$
6 $\frac{4}{6}$ and $\frac{2}{6}$

SET 4

1 40
2 600
3 3
4 6
5 4 hund, 0 tens, 7 ones
6 321
7 91
8 50c and 20c
9 $1.05
10 7
11 3, 2
12 3, 4
13 5, 10
14 4, 9
15 5, 8

Number and Algebra

1	52 – 16	becomes	56	–	20	=	36
2	46 – 28	becomes	48	–	30	=	18
3	32 – 14	becomes	38	–	20	=	18
4	163 – 57	becomes	166	–	60	=	106
5	194 – 77	becomes	197	–	80	=	117

6 30
7 50
8 60

Space

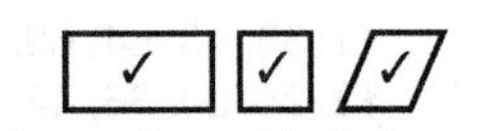

Quadrilaterals are four-sided shapes.

UNIT 31 Number and Algebra

SET 1

1 8
2 22
3 40
4 6
5 18
6 42
7 24
8 45
9 24
10 25
11 96
12 9
13 17
14 6
15 876

SET 2

1 1000
2 150
3 4006
4 7 tens
5 6 thousands
6 1234, 1341, 2341
7 653
8 149
9 4511
10 Hands on; 6509

SET 3

1 10, 13, 16, 19, 22, 25, 28
2 33, 37, 41, 45, 49, 53, 57
3 66, 71, 76, 81, 86, 91, 96
4 23, 29, 35, 41, 47, 53, 59
5 55, 60, 65, 70
6 60, 55, 50, 45
7 73, 68, 63, 58

SET 4

1 734
2 707
3 227
4 28
5 103
6 $9.07
7 7 r 1
8 32
9 $27
10 60
11 Yes
12 $7.30
13 $1.50
14 391
15 Feathers
16 Yes

Space

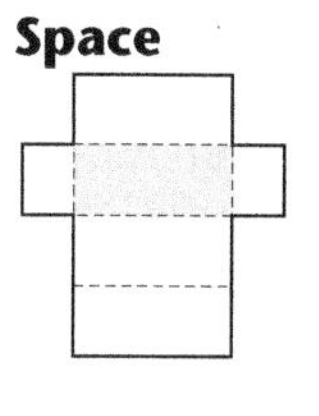

Measurement

Hands on

UNIT 32 Number and Algebra

SET 1

1 38
2 21
3 13
4 18
5 5
6 15
7 32
8 19
9 41
10 33
11 42
12 10
13 25
14 21
15 19

SET 2

1 13, 15, 17, 19, 21, 23, 25, 27
2 5, 15, 25, 35, 45, 55, 65, 75
3 90, 85, 80, 75, 70, 65, 60, 55
4 $\frac{10}{10}$, $\frac{9}{10}$, $\frac{8}{10}$, $\frac{7}{10}$, $\frac{6}{10}$, $\frac{5}{10}$, $\frac{4}{10}$, $\frac{3}{10}$
5 48, 42, 36, 30, 24, 18, 12, 6
6 6, 12, 18, 24, 30, 36, 42, 48
7 Hands on
8 Hands on
9 Hands on

SET 3

1 28
2 37
3 26
4 55
5 39
6 15

SET 4

1 20
2 8000
3 520
4 4
5 $3
6 3000
7 397
8 163
9 326
10 $3.78
11 $18
12 3580
13 300
14 45c
15 3
16 475

Space

Name	Trapezium	Hexagon	Parallelogram	Pentagon	Triangle
Vertices	4	6	4	5	3

Measurement

1 12 cm^3
2 6 cm^3
3 16 cm^3
4 36 cm^3

UNIT 33 Number and Algebra

SET 1

1 22
2 90c
3 36
4 5 tens, 7 ones
5 18
6 16
7 3
8 1999
9 One hundred and fifty-two
10 347
11 172
12 50c
13 16
14 36
15 1357

SET 2

1 3316
2 2226
3 5138
4 1409
5 2248
6 2235
7 $1377

SET 3

1 18, 18, 18
2 18, 18, 18
3 36, 36, 36
4 10, 10, 10
5 No
6 No

SET 4

1 17
2 7000
3 4 × 5
4 $\frac{4}{10}$
5 1141
6 150
7 0
8 1250
9 458
10 32
11 4 r 1
12 28
13 367 + 582 = 949 Hands on

Number and Algebra

1 4 hundreds 1 tens 8 ones
41 tens 8 ones

2 5 hundreds 7 tens 9 ones
57 tens 9 ones
579 ones

3 16 tens
4 24 tens
5 33 tens
6 58 tens
7 18 hundreds
8 26 hundreds
9 58 hundreds
10 42 hundreds

Measurement

1 6
2 2
3 4
4 1
5 3
6 5

Answers

UNIT 34 Number and Algebra

SET 1

1 $80
2 2004
3 40
4 30
5 8
6 29
7 25
8 32
9 35
10 47
11 $30
12 $15
13 5569

SET 2

1 0.32
2 0.68
3 0.15
4 0.27
5 0.45
6 0.17
7 0.64
8 0.75
9 0.81
10 0.19, 0.38, 0.58, 0.72

SET 3

1 12, 18, 15, 21, 27, 24
2 15, 35, 30, 40, 45, 25
3 6, 12, 24, 54, 36, 30
4 9
5 30
6 12
7 20
8 24
9 18
10 35
11 42

The winning card is **a**.

SET 4

1 20c each
2 30
3 22
4 300
5 $8.50
6 Two hundred and sixty-seven
7 3 books
8 1336
9 20
10 $\frac{1}{3}$
11 306, 309
12 448
13 6 r 3
14 $24
15 $28
16 $20

Space

Hands on

Measurement

Bed (2 m)
Broom (150 cm)
Pencil (14 cm)
Car (4 m)

UNIT 35 Number and Algebra

SET 1

1 17
2 19
3 12
4 9
5 40
6 30
7 21
8 16
9 61
10 15
11 30
12 120
13 22
14 12
15 18

SET 2

1 266 km
2 313 km
3 283 km
4 292 km
5 405 km
6 281
7 $173
8 105

SET 3

1 15
2 17
3 ×
4 –
5 38
6–11 Hands on

SET 4

1 842
2 35
3 $20
4 24
5 32
6 476
7 527
8 400
9 27
10 6
11 $40
12 13
13 Seven hundred and ninety-seven
14 220
15 240
16 180
17 $5

Space

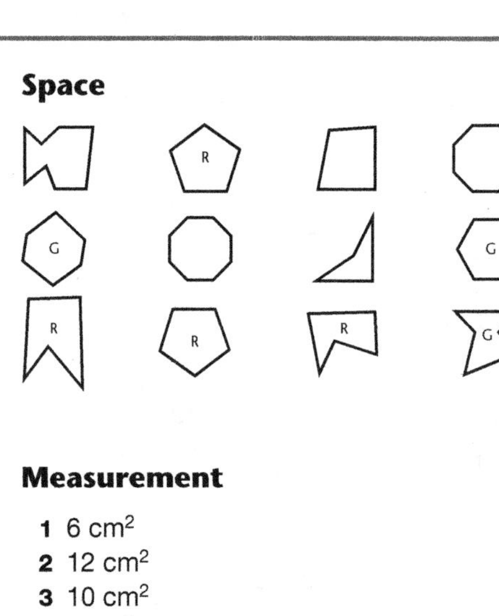

Measurement

1 6 cm^2
2 12 cm^2
3 10 cm^2
4 6 cm^2